Tobias Rauch

Spectral Enhancement of Organic Photodetectors

Spectral Enhancement of Organic Photodetectors

by
Tobias Rauch

KIT Scientific Publishing

Dissertation, Karlsruher Institut für Technologie (KIT)
Fakultät für Elektrotechnik und Informationstechnik
Tag der mündlichen Prüfung: 07. Juli 2013
Referent: Prof. Dr. Uli Lemmer
Korreferent: Prof. Dr. Wolfgang Heiss

Impressum

KIT Scientific Publishing

Karlsruher Institut für Technologie (KIT)
KIT Scientific Publishing
Straße am Forum 2
D-76131 Karlsruhe

KIT Scientific Publishing is a registered trademark of Karlsruhe Institute of Technology. Reprint using the book cover is not allowed.

www.ksp.kit.edu

Print on Demand 2014

ISBN 978-3-7315-0154-1
DOI: 10.5445/KSP/1000037753

Spectral Enhancement of Organic Photodetectors

Zur Erlangung des akademischen Grades eines
Doktors der Ingenieurwissenschaften
(Dr.-Ing.)

bei der Fakultät für Elektrotechnik und Informationstechnik
des Karlsruher Instituts für Technologie (KIT)

genehmigte
DISSERTATION

von

Dipl.-Ing. Tobias Rauch

geb. in Landsberg am Lech

Datum der mündlichen Prüfung:	07. Juli 2013
Referent:	Prof. Dr. Uli Lemmer
Korreferent:	Prof. Dr. Wolfgang Heiss

Für Mona

Für Theresa und Philipp

Abstract

Organic devices offer manifold opportunities in device processing, applied materials and system integration. Up to now, research activities have concentrated mainly on photovoltaic applications and light emitting diode in the visible spectrum. Major intention of this work is to find new alternative approaches to extend the spectral sensitivity of organic photodetectors, especially to the near-infrared regime.

In recent years, colloidal quantum dots became a promising alternative in optical and optoelectronic applications as absorbers or emitters. Similar properties of colloidal quantum dots like solution processing and engineering of the absorption characteristics make them a very interesting candidate for embedding them into organic devices. Low bandgap semiconductors (e.g. lead sulfide, mercury telluride, lead selenite, etc.) are widely used for the fabrication of colloidal quantum dots and therefore cover a broad range of the near-infrared spectrum.

In this work, embedding colloidal quantum dots of lead sulfide into organic layers shows to be a suitable approach for near-infrared detection. A major progress has been the application of the quantum dots as photosensitizer and the organic semiconductor as charge carrier transporting material in photodetecting devices. These composites of quantum dots and the organic bulk heterojunction (here: polythiophene / fullerene) demonstrated superior characteristics as high efficiencies (EQE of more than 50% at 1200 nm at a bias of -8 V) and long lifetimes with nearly preserving the properties of organic photodetectors in leakage currents and dynamics. For these inorganic-organic hybrid photodetectors, an approach for the design is derived and analyzed by intensive electro-optical characterization.

A highlight of this work is the successful integration of the hybrid devices to an a-Si active matrix thin film transistor flat panel for imaging application. An high resolution movie with an illumination source at a wavelength of 1350 nm has been acquired. Movements of the insects and their details are clearly distinguishable.

Furthermore, the detection of high energy radiation (X-rays) are studied for hybrid photodetectors since lead is a good x-ray absorber. In a proof-of-concept study, signal detection is demonstrated and an outlook for the optimization of the material properties for efficient x-ray detection is given.

Last but not least, a near-infrared absorbing polymer (here:PCPDTBT) is investigated as well. By blending together with a polymer absorbing in the visible (here: P3HT) a tuning of the total spectral sensitivity up to wavelengths of 900 nm was achieved.

Zusammenfassung

Organische Bauelemente bieten vielfältige Möglichkeiten bei den Herstellungsverfahren, den verwendbaren Materialien und Werkstoffen sowie in der Systemintegration. Bisher haben sich in diesem Bereich die Forschungsaktivitäten hauptsächlich auf photovoltaische Anwendungen und Licht emittierende Dioden im sichtbaren Bereich beschränkt. Diese Arbeit beschäftigt sich mit Ansätzen zur Erweiterung der spektralen Empfindlichkeit von organischen Photodetektoren, vor allem für den nahinfraroten Bereich.

In den letzten Jahren entwickelten sich kolloide Quantenpunkte zu einer vielversprechenden Alternative für optoelektronische Anwendung als neue Emitter- oder Absorbermaterialien. Vergleichbare Eigenschaften zu organischen Halbleitermaterialien wie lösungsbasierten Herstellungsverfahren oder gezielte Modifikation von Absorptionscharakteristiken machen die Einbettung von kolloiden Quantenpunkten in organische Bauelemente besonders interessant. Des Weiteren bestehen viele kolloide Quantenpunkte aus Halbleitermaterialien mit kleinen Bandlücken wie Bleisulfid, Bleiselenid oder Quecksilbertellurid, wodurch für optoelektronische Anwendungen weit in den nahinfraroten Bereich vorgedrungen werden kann.

In dieser Arbeit wird gezeigt, dass die Einbettung der kolloiden Quantenpunkte in organische Schichten ein wirkungsvoller Ansatz zur Erweiterung der spektralen Empfindlichkeiten von organischen Photodetektoren ist. Kolloide Quantenpunkte werden als "Photosensitizerïn eine organische Schicht, einer sogenannte "Bulk Heterojunction"(hier: Derivate von Polythiophen / Fulleren) eingebracht, die wiederum den Transport und die Extraktion der erzeugten Ladungsträger übernimmt. So erreichen diese anorganisch-organischen hybriden Photodioden mit Bleisulfid-Quantenpunkten, eine spektrale Quanteneffizienz von über 50% bei einer Wellenlänge von 1200 nm und einer Vorspannung von -8 V. Hierbei können die Eigenschaften von organischen Photodetektoren bei Leckströmen, dynamischen Verhalten und Lebensdauern annähernd beibehalten werden. Dies wird in intensiven elektrooptischen Charakterisierungen der hergestellten hybriden Bauelemente von unter anderem Stromspannungskennlinien, spektralen Empfindlichkeiten, dynamischen Verhalten, Ladungsträgertransport und Lebensdauertests nachgewiesen.

Ein Höhepunkt dieser Arbeit ist die erfolgreiche Integration der hybriden Photodioden in einen Bildsensor. Hierzu werden die Photodioden auf ein Substrat aus amorphem Silizium mit einer Aktivmatrix-Pixelansteuerung aus Dünnschichttransistoren abgeschieden. Ein hochauflösender Film mit einer Beleuchtungsquelle bei einer Wellenlänge von 1350 nm aufgenommen und rekonstruiert werden. In diesem Film sind Kellerasseln zu sehen, deren Bewegungen und feine Extremitäten gut zu erkennen sind.

Nachdem Bleisulfid-Quantenpunkte gute Ergebnisse bei der Erweiterung der spektralen Empfindlichkeit von organischen Photodioden gezeigt haben und Blei ein guter Absorber von Röntgenstrahlung ist, ist der gleiche Aufbau für die Detektion von Röntgenstrahlung untersucht worden. In einer Konzeptstudie konnte der Ansatz der Direktkonversion (direkte Erzeugung von Ladungsträgern durch Röntgenstrahlung) über Bleisulfid-Quantenpunkte sowie der Quasi-Direktkonversion (Konversion von Röntgenstrahlung in sichtbare Strahlung) von Gadoliniumoxysulfid -Partikeln in der organischen Schicht nachgewiesen werden. Darüber hinaus wird durch Modellierung versucht, einen Ausblick für die Optimierung der Material- und Schichteigenschaften zu geben.

Ferner wird die Verwendung eines nahinfrarot absorbierendes Polymer (hier: PCPDTBT) in organischen Photodioden analysiert. Durch Vermischung mit einem Polymer, das im Sichtbaren absorbiert (hier: P3HT), kann eine Abstimmung der spektralen Empfindlichkeit bis zu Wellenlängen von 900 nm für Detektoren vorgenommen werden.

Contents

1 Introduction 1
- 1.1 Motivation 1
- 1.2 Scope and Objective 2
- 1.3 Outline 2

2 Theoretical Background 5
- 2.1 Inorganic Semiconductors 5
  - 2.1.1 Energy Band Structures 6
  - 2.1.2 Charge Transport in Inorganic Semiconductors 10
  - 2.1.3 Generation and Recombination Processes in Semiconductors 15
  - 2.1.4 Photoconductivity in Semiconductors 18
  - 2.1.5 Junctions of Semiconductors 19
- 2.2 Colloidal Quantum Dots of Lead-Sulfide 25
  - 2.2.1 Electro-optical Properties of QDs 26
  - 2.2.2 Synthesis 29
- 2.3 Organic Semiconductor Devices 29
  - 2.3.1 Conjugated Polymers 30
  - 2.3.2 Organic Bulk Heterojunction 39
  - 2.3.3 Organic Photodetectors 41
- 2.4 Hybrid Organic/Inorganic Photodetectors with Colloidal quantum dots (QDs) 44

3 Measurement and Characterization Methods 47
- 3.1 Device Properties 47
  - 3.1.1 Electro-optical Characterization of Device Properties 47
  - 3.1.2 Transient Measurements 52
  - 3.1.3 Imaging 58
  - 3.1.4 Lifetime Analysis 62
- 3.2 Structural Characterization Methods 62
  - 3.2.1 Scanning Electron Microscopy 63
  - 3.2.2 Transmission Electron Microscopy 63

3.3 Characterization Methods for Material and Blend Properties . . . 65
3.3.1 Bulk Mobilities by Charge Extraction by Linear Increasing Voltage . . . 65
3.3.2 Field Effect Mobilities . . . 67
3.3.3 Temperature Induced Conductivity . . . 70
3.3.4 Transmission Spectroscopy . . . 70

4 Materials, Sample Topology and Processing Technologies 71
4.1 Materials Used . . . 71
4.1.1 Photoactive Layer Materials . . . 71
4.1.2 Materials for HTLs . . . 74
4.1.3 Materials for Electrodes . . . 75
4.2 Manufacturing Technologies . . . 77
4.2.1 Wet Chemical Deposition and Structuring . . . 77
4.2.2 Physical Vapor Deposition Technologies . . . 81
4.3 Preparation Methods by Device Layout . . . 84
4.3.1 Bottom Electrode . . . 84
4.3.2 Hole Transporting Layer . . . 85
4.3.3 Photoactive Layer . . . 85
4.3.4 Electron Transporting Layer . . . 85
4.3.5 Top Electrode . . . 85
4.3.6 Encapsulation . . . 86

5 Polymer Photodiodes for NIR Detectors 87
5.1 Donor/Acceptor Ratio in PCDTBT:PCBM Films . . . 90
5.2 Deposition Techniques and Post-Annealing . . . 92
5.3 Tuning the Absorption Spectra . . . 94
5.4 Transient Characteristics of NIR Polymer Diodes . . . 101

6 Design of Hybrid Photodiodes 107
6.1 Optimizing the Photoactive Layer . . . 107
6.1.1 Composition of the Photoactive Layer . . . 108
6.1.2 Concentration Dependence of PbS-QDs on Photodiode Performance . . . 113
6.2 Interlayers . . . 115
6.3 Inorganic-Organic / Metal Interfaces . . . 117
6.4 Deposition Technologies . . . 120
6.5 Influence of PbS-QD Properties on IO-HPDs . . . 122
6.5.1 Variation of the PbS-QD Diameter . . . 122
6.5.2 Blends with Various PbS-QD Diameters . . . 123
6.5.3 Ligand Shells . . . 124
6.5.4 Oxidation of PbS-QDs . . . 125
6.6 State of the Art IO-HPDs . . . 128

7 Characterization of IO-HPDs **131**
7.1 Structural Analysis by SEM and TEM Methods . . . 131
7.2 Charge Transport and Transfer Mechanisms in hybrid Films . . . 134
7.2.1 Mobilities in Hybrid Films . . . 134
7.2.2 Luminescence of Hybrid Films . . . 141
7.2.3 Spatially Resolved Photocurrents in Hybrid Films . . . 141
7.3 Electro-optical Characterization of hybrid Photodiodes . . . 142
7.3.1 Static Device Properties of IO-HPDs . . . 142
7.3.2 Dynamic Device Properties of IO-HPDs . . . 149
7.3.3 Stability of IO-HPDs . . . 157

8 PbS-QD NIR-Imager **159**
8.1 Integration of IO-HPDs on a-Si TFT Imaging Sensors . . . 159
8.2 Near-Infrared Imaging . . . 162

9 X-ray Detection with IO-HPDs **167**
9.1 Conventional X-ray Detection with Flat Panel Detectors . . . 167
9.1.1 Flat Detectors with Indirect X-ray Conversion . . . 167
9.1.2 Flat Detectors with Direct X-ray Conversion . . . 168
9.1.3 Organic Photodetectors for X-ray Detection . . . 170
9.2 Direct and Quasi-Direct Conversions in IO-HPDs . . . 170
9.2.1 Quasi Direct Conversion with Embedded Scintillator Particles . . . 174
9.2.2 Direct Conversion with Embedded Lead Sulfide Quantum Dots . . . 177
9.2.3 Comparison of Quasi-Direct and Direct Conversion Devices . . . 180

10 Outlook **185**

Appendix **189**

A Symbols and Abbrevations **191**
A.1 Symbols . . . 191
A.1.1 English Symbols . . . 191
A.1.2 Greek Symbols . . . 193
A.1.3 Physical Constants . . . 194
A.2 Abbrevations . . . 194

List of Figures **202**

List of Tables **203**

Bibliography 203

Acknowledgements 223

1. Introduction

1.1. Motivation

Nobel-Prize winners Heeger,MacDiarmid and Shirakawa created with their discovery of semiconducting conjugated polymers [18] a new field of electronics - the organic electronics. This field has been rapidly growing and progressing over the last two decades. Devices and applications have been developed and have already reached technological maturity. Material research is providing a growing range of organic materials for different applications.
For thin film transistors (TFTs) p- and n-semiconducting materials with mobilities of 10^{-2} cm^2(Vs)$^{-1}$ and 10^{-1} cm^2(Vs)$^{-1}$ exceed the mobilities of amorphous silicon (a-Si) [79] and have reached a value needed for applications such as organic radio frequency identifications (RFIDs). Organic photovoltaics has continuously increased the performance, achieves power conversion efficiency up to 12% and meets standard lifetime requirements of conventional photovoltaics (see `http://www.heliatek.de`, (May 28, 2013)). But by far the largest growing sector of organic electronics is related to organic light emitting diodes (OLEDs) for display and lighting applications. Samsung demonstrated a 55-inch TV on the International Consumer Electronic Show in Las Vegas on January 11th, 2012 (see `http://www.samsung.com/us/news`). This strong technological progress in these fields of organic electronics is reflected by different optimistic forecasts for growing markets from 8 billion US-dollars in 2012 to 45 billion in 2018. Especially new applications will drive the market (see `http://www.wallstreet-online.de` *Organic Electronics Market Will Reach USD 44.8 Billion Globally in 2018* on April 10, 2012).
These successful stories are clearing the way for further organic electronic devices for special applications. In particular, organic photodetectors (OPDs) have demonstrated high potential for implementation in optoelectronic systems. Suitable properties for tuning spectral sensitivity, dynamics and large area processing made OPDs a promising candidate for imaging and sensor applications. In particular, new approaches for inorganic-organic hybrid devices broadened the application opportunities of organic optoelectronic devices [139]. Organic bulk heterojunctions are attractive as solution-processable composites, but are limited to use in the visible spectrum. According to current

progress in near-infrared (NIR) imaging, cost-effective devices are highly demanded.

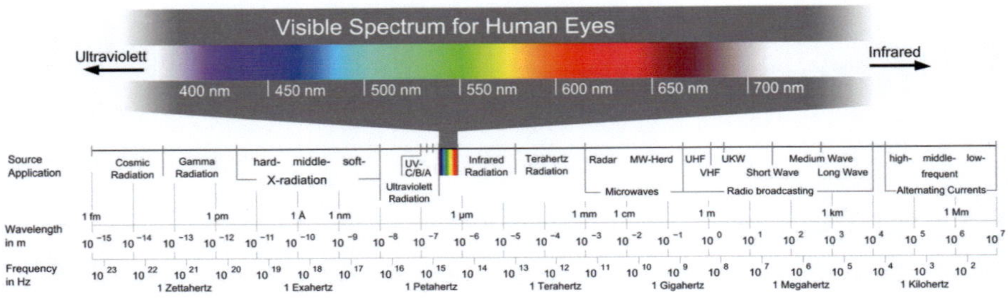

Figure 1.1.: The Spectrum of Electromagnetic Radiation. The very broad spectrum of electromagnetic radiation is only visible in a tiny range human eyes, but widely used by electronic and optoelectronic applications. This figure was modified from `http://de.wikipedia.org/wiki/Elektromagnetisches_Spektrum` (May 25, 2012).

1.2. Scope and Objective

The objective of this thesis is the development and analysis of possibilities to extend the photoresponse of organic photodetector and organic imaging devices to the NIR regime by preserving the advantages of soluble processing organic polymers under the perspective of industrial applications. Focus of the work is the modification and investigation of the photoactive layer by newly defined or adjusted composites for NIR sensitivity and their application in photodetection devices. Since there were only few polymers reported for NIR responsivity, this work for OPDs was not restricted to conjugated organic polymers only, but includes also inorganic semiconductors for a hybrid approach. The optimization and integration of promising composites for sensing and imaging applications will be considered as well in this work. Furthermore, as the investigated materials of colloidal QDs are not only absorbing in the NIR, these materials will be additionally considered for detection of high-energy radiation.

1.3. Outline

The theoretical background of inorganic, organic semiconductor and quantum dots with reference to optoelectronic applications is given in chapter 2. Applied

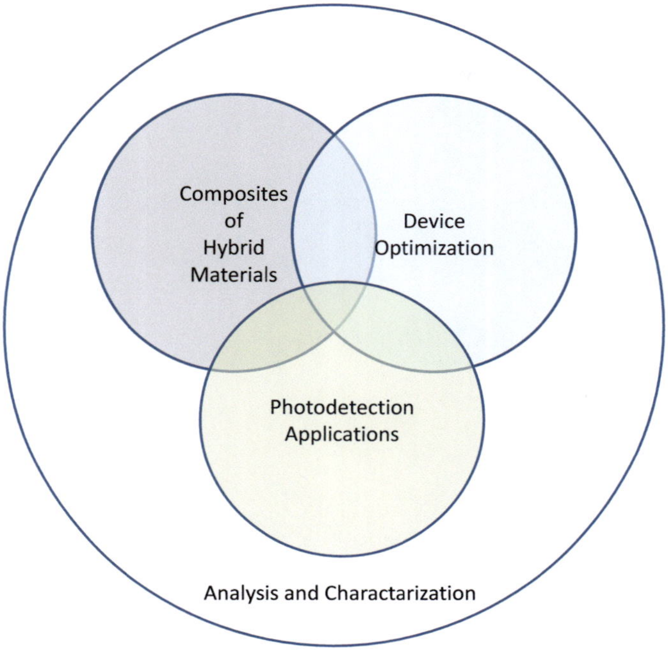

Figure 1.2.: Major activities for extending the spectral sensitivity of OPDs to the NIR.

methods for analysis and characterization of samples, composites and applications within this work are introduced in chapter 3. A detailed overview on processing technologies, applied process conditions and further treatments of materials and samples is described in chapter 4. In chapter 5, the modification and the shift of the response to the NIR by substitutes to the current polymer is investigated. The following three chapters are dedicated to an approach for inorganic-organic hybrid photodiodes (IO-HPDs) using colloidal QDs as sensitizers. At first, in chapter 6 the composite and photodiode themselves are explored for the purpose of device design. Subsequently, the optoelectronic properties and characteristics of IO-HPDs are investigated in chapter 7. Finally in chapter 8, IO-HPDs are integrated on active matrix TFTs backplanes to demonstrate NIRs imaging with the introduced hybrid approach on the system level. Chapter 9 is assigned for research on high energy particle detection like x-rays as the QD material of lead sulfide is known as a strong x-ray absorber.

2. Theoretical Background

In this work optoelectronic organic and hybrid inorganic/organic devices are modified for near-infrared detection. As inorganic and organic semiconductor play a major role in these devices the fundamentals for understanding the principles are dealt within this chapter.

2.1. Inorganic Semiconductors

Semiconductors are an own kind of material and are neither conductors nor insulators. This can be explained by figure 2.1 where the energy band diagram (discussed later in this chapter) for each type is shown. Conductors have overlapping valence and conduction bands (e.g. Pb) or partially filled conduction bands (e.g. Cu) yielding typical conductivities of more than $10^6 Scm^{-1}$. In both cases no energetic barriers need to be overcome for electrons (negative charge carrier) and holes (positive charge carrier: defect electron). An insulator exhibits a very large gap ($\gg$ 3 eV) of the conduction and valence band prohibiting the generation of charge carriers, so that the conductivity of insulators is typically below $10^{-12} Scm^{-1}$. In contrast, semiconductors have a sufficient narrow band gap from a few hundred meV to 3 eV between the valence and conduction band to allow the generation of charge carriers, and thus intrinsic conductivity.

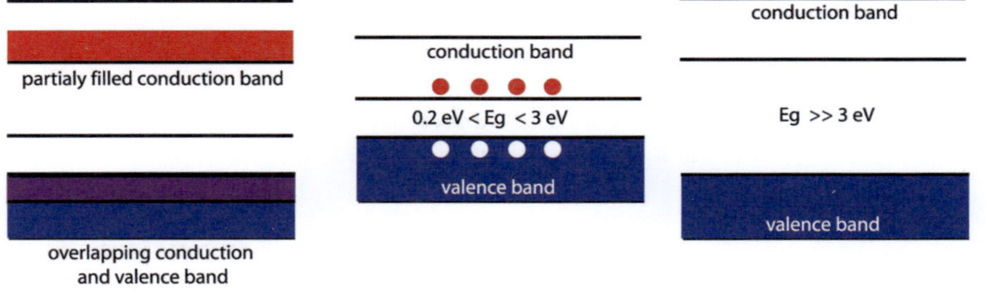

Figure 2.1.: Energy band gaps of conductors, semiconductors and insulators

2.1.1. Energy Band Structures

The energy band structures belong to the effect, that an electron within condensed matter experiences different potentials according to the arrangement of atoms. The electron is not free to travel and has to obey allowed energetic states. For a better understanding of the energetic structure, a one-dimensional quantum well is discussed as an approximation for an atom. This is then extended within the Kronig-Penney model to consider the band structures in ordered materials as crystals or polymers.

An electron in a one dimensional quantum well

The particle/wave dualism in quantum mechanics allows describing electrons by a complex wave equation depending on time and space called Schrdinger equation (eq. 2.2). With the solution of the time-independent Schrdinger equation (eg. 2.3) one is able to determine steady state of energy levels and the probability density of the electrons in an arbitrary potential profile. The Schrdinger equation is given by

$$i\hbar\frac{\partial\Psi}{\partial t} = -\left(\frac{\hbar}{2m}\nabla^2 + V(t)\right)\Psi \tag{2.1}$$

and the the eigenfunction for the Schrdinger equation can be derived:

$$\Psi(\vec{r},t) = \Psi(\vec{r})\cdot exp\left(\frac{-iEt}{\hbar}\right) \tag{2.2}$$

with $\Psi(\vec{r})$. The eigenfunction of time-independent Schrdinger equation can be expressed:

$$\nabla^2\Psi + \frac{2m_e}{\hbar^2}(E-V)\Psi = 0. \tag{2.3}$$

Here, E is the Eigenvalue of the energy, $\hbar = h/2\pi$ is the reduced Planck's constant, m_e the electron mass and V the actual potential.
Here, we have a look at a quantum well with infinite barriers (see figure 2.2) which is a one-dimensional structure so that we consider only the the solution for plane waves in a one dimensional system. We can reduce the Schrdinger then to the following:

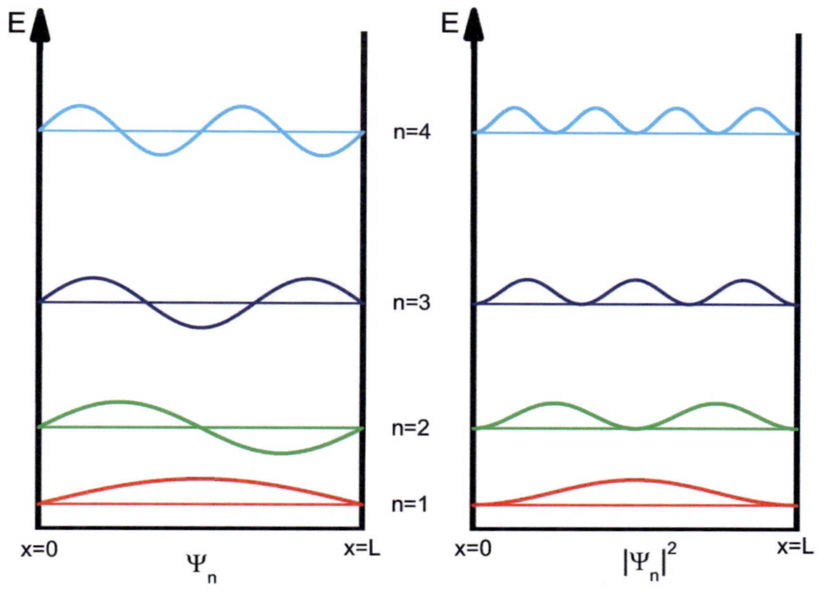

Figure 2.2.: Solution of the Schrdinger equation in a quantum well with infinite barriers. Discrete energy levels with the wave function (**left**) and with the probability density for the electron (**right**).

$$\frac{\partial^2}{\partial x^2}\Psi(x) + k^2\Psi(x) = 0 \tag{2.4}$$

with the wavenumber

$$k = \sqrt{\frac{2m_e E}{\hbar^2}} \tag{2.5}$$

As shown in figure 2.2 the potential profile is $V(0 < x < L) = 0$, $V(x < 0) = \infty$ and $V(x > L) = \infty$ giving the boundary conditions:

$$\Psi(0) = \Psi(L) = 0 \tag{2.6}$$

The Schrdinger equation is solved with the ansatz:

$$\Psi(x) = Asin(kx) + Bsin(kx) = 0 \tag{2.7}$$

Considering the boundary conditions yields

$$\Psi(0) = B = 0 \tag{2.8}$$

and

$$\Psi(L) = Asin(kL) = 0 \Rightarrow k = \frac{n\pi}{L} \text{ with } n \in Z \setminus \{0\} \tag{2.9}$$

With equation 2.5 the energy levels are given by

$$k = \frac{n^2\pi^2\hbar^2}{2m_e L^2} \tag{2.10}$$

For these reasons the electrons can only occupy discrete energy values. The energy values for $n = 1 \ldots 4$ are depicted in figure 2.2. The Schrdinger wave function $\Psi(x)$ is illustrated in figure 2.2 on the left and the probability density $|\Psi(x)|^2$ on the right.

The Kronig-Penney model

We have to deal with materials of high periodicity. The lattice constant gives the dimensions of a primary cell which is periodically repeated within the crystal. Thus, in all of the three dimensions a periodic configuration of the atoms is found. In the Kronig-Penny model these atoms are seen as positively charge ions creating a periodic 1D potential. In figure 2.3 the simplification from the potential profile of positive ions to a system of coupled quantum wells can be seen.

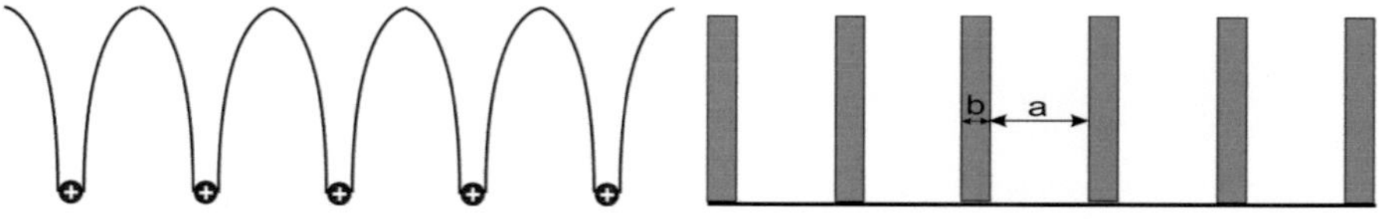

(a) 1D periodic potential of positive ions (b) Simplification by Kronig-Penney model

Figure 2.3.: The Kronig Penney model: The periodicity of the potential in a one-dimensional crystal can be expressed in an approximation by coupled quantum wells.

Due to the periodicity of the Kronig-Penney model the Bloch theorem can be introduced to simplify the solution of the Schrdinger equation. The Bloch theorem for a periodic 1D potential is

$$V(x+T) = V(x) \tag{2.11}$$

with T for the period. Thus, for a 1D periodic potential the Schrdinger equation fulfills with the Bloch theorem

$$\Psi(x+T) = exp\,(ikT) \cdot \Psi(x) \tag{2.12}$$

so that the solution needs to be found only for a single period analog to a quantum well. This yields to the conclusion that for each possible energy value two different k-values are possible ($\pm k$). Moreover, for k values where $k \pm n \cdot 2\pi/a, n = 1, 2, 3, \ldots$ the same energy value and wave functions can be assigned. Therefore the energy values can be plotted in a reduced manner in the first Brillouin zone from $-\pi/a$ to π/a.

As can be seen in figure 2.4, there are some regions of prohibited energy values because of the discontinuity at the boarders of the Brillouin zone yielding to energy band gaps (E_g). The dimension of the gap determines the semiconducting or insulating properties.

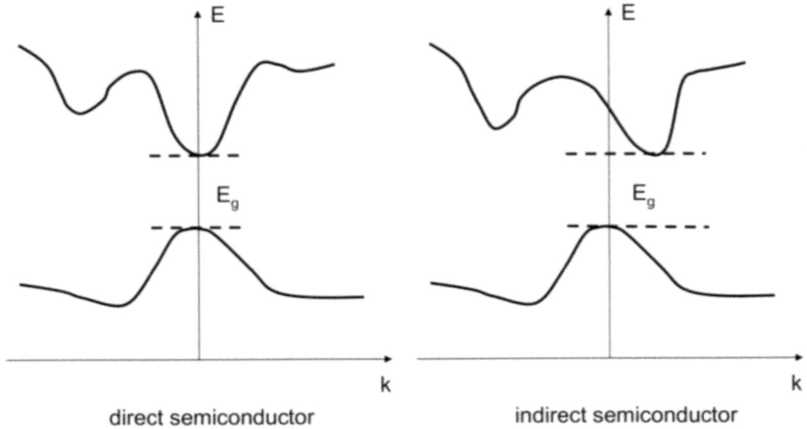

Figure 2.4.: Energy band gaps of semiconductors with direct and indirect transitions

The presentation of the theory relies mainly on ref. [164, 53].

2.1.2. Charge Transport in Inorganic Semiconductors

Now, let's consider not only the energy conditions of particles, but as well the properties for charge transport in solid state. Transport is performed by electrons in the conduction band and by holes (defect electrons) in the valence band.

Mobility and effective masses

The mobility of charge carriers is particularly determining the transport in semiconductors. If a charge carrier experiences a force (e.g. an electric field), the carrier is forced to move with its mobility accordingly to the force. The classical transport model of Drude defines the mobility as

$$\vec{v_D} = \mu \vec{E}, \quad \text{with} \tag{2.13}$$

$$\mu = \frac{q t_c}{m} \tag{2.14}$$

$\vec{v_D}$ is the drift velocity, $\vec{E}$ the electric field and μ the mobility. Furthermore, the mobility μ depends on the charge of the species q, the mean time between scattering or collision events t_c and and it depends inverely on the mass m.

In order to account for the periodic potential in a semiconductor crystal, the charge carrier cannot be considered as a free electron. Therefore an effective mass is introduced. The effective mass is derived for a 1D case from the first Brillouin zone of the conduction band minimum and valence band maximum

$$\mu = \frac{qt_c}{m^*}, \text{ with} \tag{2.15}$$

$$m^* = \left(\frac{\partial^2 E(k)}{\partial k^2}\right)^{-1} \cdot \hbar^2 \tag{2.16}$$

Note that this approximation is only constant for region where $E \propto k^2$.

Conductivity and Doping

The conductivity σ is defined as the proportionality factor in the relation between the applied field $\vec{E}$ and the current density $\vec{J}$ and expresses the ability to conduct current. Obviously, the mobility and densities of available charge carrier is as well of importance for the conductivity. The conductivity of a material is therefore given by

$$\sigma = qn\mu_n + qp\mu_p \tag{2.17}$$

with the electron densities n, the hole densities p the electron mobility μ_n and hole mobility μ_p.

The carrier densities in thermal equilibrium of an intrinsic semiconductor at a certain temperature follow the Fermi-Dirac distribution which gives the probability for the occupancy of states.

$$f(E) = \frac{1}{1 + e^{(E-E_F)/k_B T}} \tag{2.18}$$

E_F is the Fermi energy, k_B is the Boltzmann constant and T the temperature. The Fermi energy is the value where half of the states are occupied. Integrating the product of Fermi-Dirac distribution and the density of states for electrons allows the determination of the carrier densities n in the conduction band and for holes p in the valence band, respectively allows calculating the carrier densities by

$$n = N_C exp\left(-\frac{E_C - E_F}{k_B T}\right) \tag{2.19}$$

$$p = N_V exp\left(-\frac{E_F - E_V}{k_B T}\right) \tag{2.20}$$

Note that equations 2.19 and 2.20 are only valid for non-degenerate semiconductor (Fermi energy lies within the energy band gap). For intrinsic semiconductors ($n = p$) the intrinsic Fermi energy E_i is

$$E_i = \frac{E_C + E_V}{2} + \frac{k_B T}{2} \ln\left(\frac{N_C}{N_V}\right). \tag{2.21}$$

Vice versa the energy band gap can be determined by

$$E_g = k_B T \ln\left(\frac{N_V N_C}{n_i^2}\right) \tag{2.22}$$

With doping the carrier densities n or p can be dramatically increased and so the conductivity. Doping atoms are embedded in the crystal to create traps which act as acceptors (p-doping) or as donors (n-doping). The energy of the traps lies close to the conduction band or to the valence band for n-doping and for p-doping, respectively (see. figure 2.5). The following charge neutrality condition can be assumed for large enough temperatures, so that all of the traps are ionized.

$$n + N_A = p + N_D, \tag{2.23}$$

with the acceptor density N_A and the donor density N_D.

(a) n-doping (b) p-doping

Figure 2.5.: Fermi energy levels due to n- and p-doping. Due to the dopants the Fermi energy is shifted either close to the conduction band or to the valence band for (**a**) n- and (**b**)p-doping, respectively

Combining equations 2.19, 2.20 and 2.23 the Fermi energies in case of doping are given by

$$E_{F,n} = E_C - k_B T \ln\left(\frac{N_C}{N_D}\right) \tag{2.24}$$

$$E_{F,p} = E_V + k_B T \ln\left(\frac{N_V}{N_A}\right) \tag{2.25}$$

The drift-diffusion model

The transport mechanism in semiconductors is mainly driven by two mechanisms: drift and diffusion. Drift of carriers is due to an electric field, whereas concentration gradients of carriers are responsible for the diffusion.

$$\vec{J}_{diff,n} = qD_n \nabla n \tag{2.26}$$

$$\vec{J}_{diff,p} = -qD_p \nabla p \tag{2.27}$$

with the diffusion coefficients for electrons D_n and for holes D_p from the Einstein relation for non-degenerate semiconductors.

$$D_n = \frac{k_B T}{q}\mu_n \tag{2.28}$$

$$D_p = \frac{k_B T}{q}\mu_p \tag{2.29}$$

As shown in figure 2.6(a) due to the one dimensional concentration gradients in the carrier density, the carriers are forced to equalize the concentration throughout the semiconductor along the x-dimension. Thus, a diffusion current density $\vec{J}_{diff}$ occurs as the charge carriers are moving. As soon as an electric field is applied (e.g. external or built-in) an additional component to the diffusion current needs to be considered, because now, the charge carriers like to travel along the applied field. Let's regard solely the drift current density with neglecting the diffusion. The drift current density $\vec{J}_{drift}$ is related with the conductivity to the

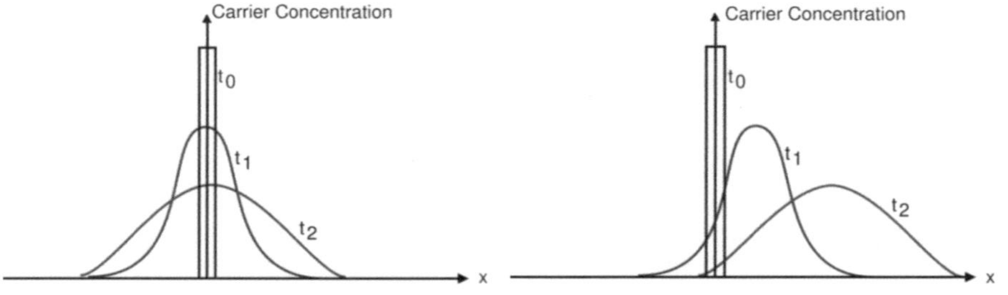

(a) Diffusion

(b) With additional drift

Figure 2.6.: One dimensional diffusion and drift of charge carriers. (**a**) Diffusion of charge carriers takes place because of their gradient in concentration and (**b**) with an applied field along the x-axes the charge carriers experience a drift.

electric field by $\vec{J}_{drift} = \sigma\vec{E}$. Applying equation 2.17, the drift current density can be split for electrons and holes respectively.

$$\vec{J}_{drift,n} = q\mu_n n\vec{E} \tag{2.30}$$

$$\vec{J}_{drift,p} = q\mu_p p\vec{E} \tag{2.31}$$

The whole current density results from the sum of the drift and diffusion component for each charge carrier.

$$\vec{J}_n = \vec{J}_{drift,n} + \vec{J}_{diff,n} = q\mu_n n\vec{E} + qD_n\nabla n \tag{2.32}$$

$$\vec{J}_p = \vec{J}_{drift,p} + \vec{J}_{diff,p} = q\mu_p p\vec{E} - qD_p\nabla p \tag{2.33}$$

As we are mostly dealing with one dimensional problems, equations 2.32 and 2.33 can be simply reduced by replacing the ∇-operator by $\partial/\partial x$. The overall current density is $\vec{J} = \vec{J}_n + \vec{J}_p$:

$$\vec{J} = q\left(\mu_n n + \mu_p p\right)\vec{E} + q\left(D_n\nabla n - D_p\nabla p\right) \tag{2.34a}$$

with the Einstein relation of equations 2.28 and 2.29

$$\vec{J} = q\left(\mu_n n + \mu_p p\right)\vec{E} + k_B T\left(\mu_n\nabla n - \mu_p\nabla p\right) \tag{2.34b}$$

The continuity equations

In semiconductors the carrier densities obey the continuity conditions. This means, that for an infinitesimal volume, the variation of carrier densities by time equals the difference of incoming and outgoing current flux plus the difference of the generation and the recombination of carriers.

$$\frac{\partial n}{\partial t} = G_n - R_n + \frac{1}{q}\nabla\vec{J}_n \tag{2.35a}$$

$$\frac{\partial p}{\partial t} = G_p - R_p - \frac{1}{q}\nabla\vec{J}_p \tag{2.35b}$$

with G and R as the generation rates for the respective charge carrier.

Ref. [164, 53] show the fundamentals in this subsection also in more details.

2.1.3. Generation and Recombination Processes in Semiconductors

As generation and recombination mechanisms are of particular interest for optoelectronic devices, a better understanding with a special focus on radiative generation and relevant recombination processes is of fundamental importance. In figure 2.7(b) generation and absorption mechanisms are indicated by 1 to 3, while recombination processes are labeled with 4 and 5. Detector characteristics strongly depend on these processes.

Recombination of charge carriers

In photodetectors, recombination processes mostly should be kept as small as possible, since generated charge carriers by radiation need to be fully extracted in an ideal photodetector before recombination. In a simple approximation, the charge carriers need to be extracted faster than their lifetime τ in the device yielding to a net recombination rate U of

$$U_n = \frac{n_p - n_{p,0}}{\tau_n} \text{ for p-type semiconductors} \tag{2.36a}$$

$$U_p = \frac{p_n - p_{n,0}}{\tau_p} \text{ for n-type semiconductors} \tag{2.36b}$$

In a quasi-neutral semiconductor the recombination depends on the excess minority charge carriers $n_p - n_{p,0}$ and $p_n - p_{n,0}$, so that in steady state the majority charge carrier recombination is similar to the minority charge carriers (an equal number of holes and electrons is necessary). Below, three major relevant recombination mechanisms for hybrid photodetectors are in focus.

Band to Band Recombination This type of recombination involves both charge carriers holes and electrons. An electron from the conduction band recombines with a hole from the valence band under releasing the inherent energy in e.g. a photon (see figure 2.7(b) #4). For this reason the recombination process depends on the number of each charge carrier. The band to band recombination can be written as

$$U_{b-b} = b\left(np - n_i^2\right) \tag{2.37}$$

with b as the bimolecular recombination constant. In photodetectors to avoid band to band recombination, charges need to be transferred faster to the extraction electrodes than their lifetime. A special and extended form of the band-to-band recombination is the Auger recombination. A electron-hole pair recombines by transferring the energy to another electron in the conduction, which relaxes by scattering to the conduction band level (a combination of #4 and #3 in figure 2.7(b)).

Trap assisted Recombination Especially in organic and hybrid photodetectors recombination induced by traps is of importance since the devices consist of different amorphous materials, so that the trap level is significantly higher compared to ordered inorganic semiconductors. As depicted in figure 2.7(b) #5 charge carriers can recombine via trap levels in the conduction band. Those trap levels can be introduced by the interaction of the different materials in organic and hybrid semiconductors and by ambient processing conditions. In ref. [171, 53] the recombination over traps is express by

$$U_t = \nu_{th}\sigma_0 N_{th} \frac{pn - n_i^2}{p_+ n + 2n_i \cosh\left(\frac{E_t - E_i}{k_B T}\right)} \tag{2.38}$$

which becomes for $n \gg p$ to

$$U_t \approx \frac{p_n - p_{n,0}}{\tau_p} \tag{2.39}$$

Surface Recombination In hybrid and organic devices, a very inhomogeneous morphology introduces many interfaces (c.f. organic bulk-heterojunctions) between the involved materials and composites. At these interfaces, a surface type recombination can be found. At surfaces or interfaces, traps form recombination centers as the morphology of the domains abruptly changes. Since traps at the interfaces are involved, the expression is very similar to the one for trap assisted recombination:

$$U_s = \nu_{th}\sigma_0 N_{st} \frac{pn - n_i^2}{p_+ n + 2n_i \cosh\left(\frac{E_i - E_{st}}{k_B T}\right)}. \tag{2.40}$$

Under the condition $n \gg p$ it can be reduced to

$$U_s \approx \nu_{th}\sigma_0 N_{st} p_n - p_{n,0}. \tag{2.41}$$

Radiative generation of charge carriers

In photodetectors, the utilization of photoinduced generation of carriers is the operating principle of these devices. With an incident photon flux (light) on

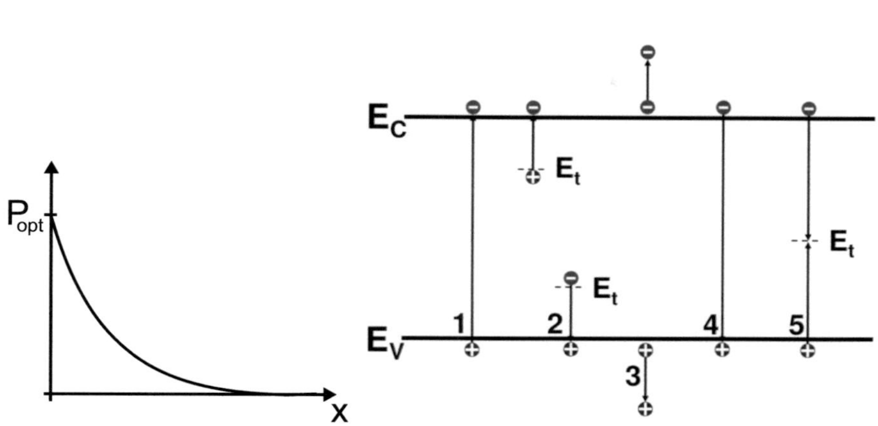

(a) Beer's Law (b) Generation and Recombination processes

Figure 2.7.: (**a**) Absorption in matter according to Beer's law. (**b**) Radiative generations and recombination of charge carriers in semiconductors.

the semiconductor, photons interact with the material and can be absorbed by a transition of an electron from the valence band to the conduction band if the photon energy is larger than the band gap. A measure for the ability of an material to absorb light is the spectral dependent absorption coefficient $\alpha(\lambda)$. As light travels through matter, the intensity of the light decays exponentially by $\alpha(\lambda)$ taking into account by Beer's law.

$$P_{opt}(x) = P_{opt,0} \cdot exp\left(-\alpha\left(\lambda\right) \cdot x\right) \tag{2.42}$$

with $P_{opt,0}$ as the optical power density penetrating into the material and $P_{opt}(x)$ as the power density after traveling the length x through the material. As mentioned above not all the absorbed photons generate free carriers. The internal quantum efficiency η_{int} considers this non-ideality. For a monochromatic optical power the generation rate of free electron/hole pairs can be written as

$$G_{n,rad}(x) = G_{p,rad}(x) = \eta_{int} \cdot \frac{1}{hf} \cdot \frac{-\partial P_{opt}(x)}{\partial x} = \frac{\alpha(\lambda)}{hf} \cdot P_{opt}(x) \tag{2.43}$$

where $G_{n,rad}$ and $G_{p,rad}$ are the radiative generation rates for electrons and holes, respectively- $-\partial P_{opt}(x)/\partial x$ accounts for the absorption of photons.

Charge carrier generation can also take place from traps, which can be considered as unintended doping (see figure 2.7(b) #2). Note that absorption of light

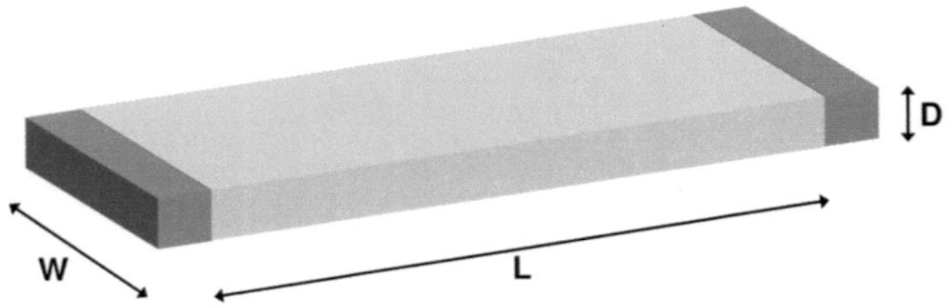

Figure 2.8.: Configuration of a semiconductor for photoconduction. The bulky semiconductor has two side contacts and exposes the large area to the light.

does not only happen by creation of an electron-hole pair, but also by exciting a charge carrier as illustrated in figure 2.7(b) #3.

For deeper information please refer again on ref. [171, 53].

2.1.4. Photoconductivity in Semiconductors

Since semiconductors absorb light and charge carriers are generated by this process, a so called photodoping and an increase in conductivity is found for semiconducting photoconductors. The rise in conductivity of a photoconductor is dependent on the generation rate G_{PC} which is given by

$$G_{PC} = \eta \cdot \frac{P_{opt}/hf}{WLD} \tag{2.44}$$

with P_{opt} as the incident optical power and the dimensions of the photoconductor (W the width, L the length, D the depth). As depicted in figure 2.8, it is assumed that the photoconductor is thick enough to completely absorb the incident light. At equilibrium the generation rate equals the recombination rate $R = n/\tau$, so that the photocurrent I_{PC} can be described by

$$I_{PC} = \underbrace{qn\left(\mu_n + \mu_p\right)}_{\sigma} \cdot F \cdot W \cdot D = \underbrace{q\eta \cdot \frac{P_{opt}}{hf}}_{I_{ph}} \cdot \frac{\left(\mu_n + \mu_p\right)\tau F}{L} \tag{2.45}$$

The primary photocurrent I_{ph} which is actually generated by the incident photons can underlie amplification by the photoconductive gain. From equation

2.45 I_{ph} is proportional to the lifetime and the transit time $t_{tr} = \frac{L}{(\mu_n + \mu_p)F}$, so that the photoconductive gain $Gain_{PC}$ is

$$Gain_{PC} = \frac{I_{PC}}{I_{ph}} = \frac{\tau}{t_{tr}} \tag{2.46}$$

Amplification of the primary photocurrent is dependent on the device geometry as small L, long lifetimes and high mobilities increase the photoconductive gain. On the contrary lifetimes are responsible for the response time of photodetectors and a compromise has to be made between gain and speed of the photoconductor [165].

A prerequisite for photoconductive gain are different mobilities of the charge carriers. The excited charge carriers with higher mobility are extracted faster at the electrodes (electrons in figure 2.9), while the slower charger carriers remain in the bulk and create a space charge as known from dopant atoms. Due to the established field and the request for charge neutrality charge carriers are injected from the electrodes. According to the mobility electrons may travel several times through the photoconductor generating a photoconductive gain [165].

2.1.5. Junctions of Semiconductors

Important configurations in semiconductor devices are the junction of semiconductors with themselves, to metals or to dissimilar semiconductors. In this subsection the different junctions are considered according to the fundamentals of semiconductor physics described in previous subsections 2.1.1, 2.1.2 and 2.1.3.

The p/n-Junction

A p/n-junction is a junction of two differently doped regions within a single semiconductor showing a rectification behavior (see figure 2.10) used in such applications as solar cells, photodiodes and light emitting diodes. For clarity and necessity abrupt p/n-junctions only are taken into account. For further readings, ref. [166] gives a broad overview of p/n-junctions.

In a p/n-junction the alignment of the Fermi-levels for the n-doped and p-doped regions creates a built-in potential (c.f. figure 2.10(a), so that a rectification can be established and can only be overcome by applying a forward

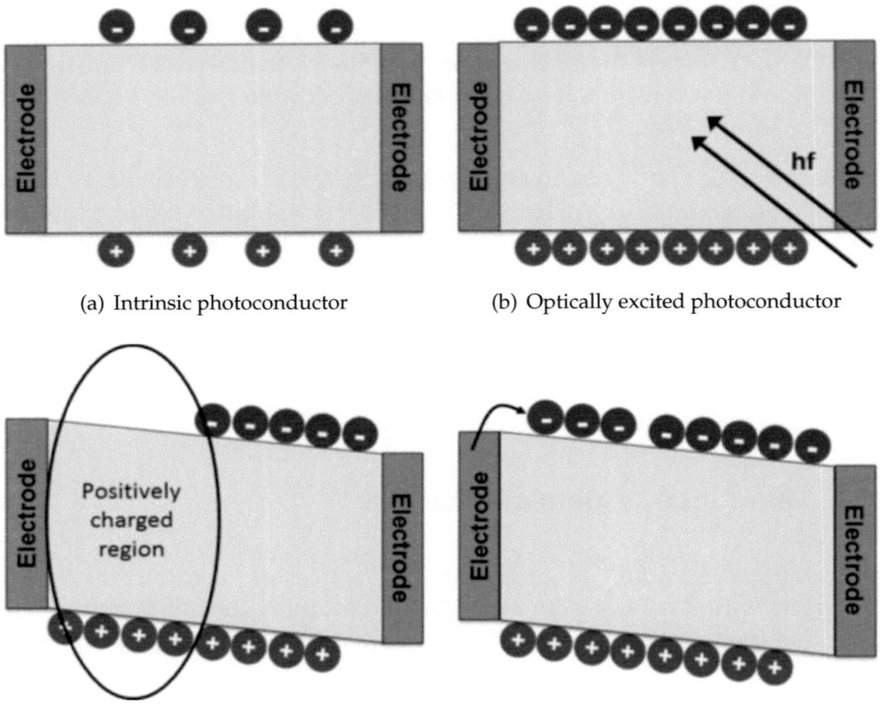

(a) Intrinsic photoconductor

(b) Optically excited photoconductor

(c) Spatially charged photoconductor

(d) Charge compensated photoconductor

Figure 2.9.: Photoconductive gain mechanism: (**a**) An intrinsic photoconductor before applying light and bias. (**b**) Additional charges are generated by incident light in the photoconductor. (**c**) Bias is applied to the photoconductor. As the electrons have a significant higher mobility, the electrons are extracted faster and the remaining holes create a positively charged region. (**d**). The holes forces electrons to be injected. Until the total decay of the excited charge carriers, electrons can travel several times through the device amplifying the primary photocurrent.

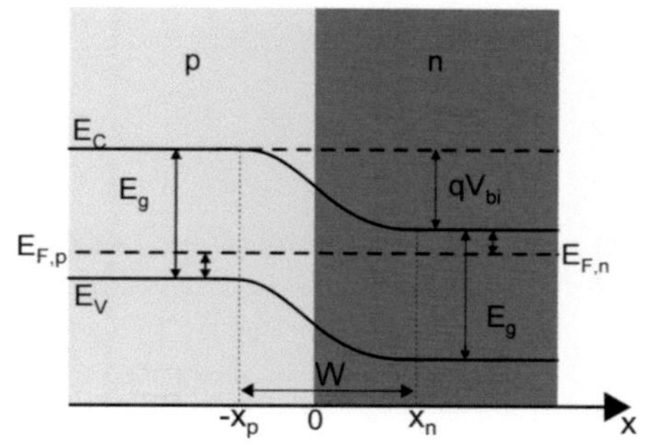

(a) pn-junction in thermal equilibrium

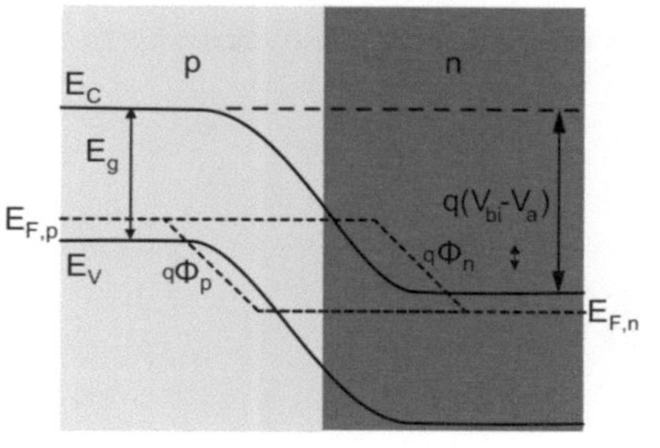

(b) pn-junction with applied reverse bias

Figure 2.10.: The pn-junction: Two different a n-doped and a p-doped region of a semiconductor are brought into contact. The distribution of the energy bands in the semiconductor are illustrated (**a**) for thermal equilibrium and (**b**) under applied reverse bias.

bias higher than the built-in voltage. The built-in potential and voltage can be determined by

$$qV_{bi} = E_g - (E_{Fp} - E_V)i(E_C - E_{Fn}) \tag{2.47}$$

E_g is the band gap, E_V the valence band edge and E_C the conduction band edge. E_{Fp} and E_{Fn} are the Fermi-levels in the p-type region and the n-type region, respectively. Considering equations 2.24 and 2.25 and that $E_C - E_V = E_g$, equation 2.47 can be written for non-degenerate semiconductors as

$$\begin{aligned} qV_{bi} &= k_BT\ln\left(\frac{N_CN_V}{n_i^2}\right) - \left(k_BT\ln\left(\frac{N_V}{N_A}\right) + k_BT\ln\left(\frac{N_C}{N_C}\right)\right) \\ V_{bi} &= \frac{k_BT}{q}\ln\left(\frac{N_AN_D}{n_i^2}\right) \end{aligned} \tag{2.48}$$

Assuming full ionization of the doping atoms (what is allowed at room temperature), $N_A \approx p_{p0}$, $N_D \approx n_{n0}$ and thermal equilibrium with the mass-action law $n_{n0}p_{n0} = n_{p0}p_{p0} = n_i^2$ leads to

$$V_{bi} = \frac{k_BT}{q}\ln\left(\frac{n_{n0}}{n_{p0}}\right) = \frac{k_BT}{q}\ln\left(\frac{p_{p0}}{p_{n0}}\right) \tag{2.49}$$

In case one region of the semiconductor is degenerate, the built-in voltage is obtained from equation 2.47.

Since a depletion layer is formed at the pn-junction, the built-in potential qV_{bi} can be also obtained by solving the Poisson equation.

$$-\frac{\partial^2 V}{\partial x^2} = \frac{\partial \overrightarrow{E}}{\partial x} = \frac{\rho(x)}{\varepsilon_s} = q\frac{p(x) - n(x) + N_A^+ - N_D^-}{\varepsilon_s} \tag{2.50}$$

$\rho(x)$ is the space charge density and ε_s the dielectric constant of the semiconductor. $n(x)$ and $p(x)$ can be neglected if full depletion is assumed for the pn-junction obtaining

$$\frac{\partial \overrightarrow{E}}{\partial x} = q\frac{N_A^+ - N_D^-}{\varepsilon_s} \tag{2.51}$$

$N_A^+ \approx 0$ in the p-doped region and $N_D^- \approx 0$ in the n-doped region yields to

$$\overrightarrow{E} = \begin{cases} -\frac{q}{\varepsilon_s} N_A(x + x_p) \\ \frac{q}{\varepsilon_s} N_D(x - x_n) \end{cases} \tag{2.52}$$

By integrating the electric field $\overrightarrow{E}$ for the space charge region, the built-in voltage V_{bi} is obtained by

$$V_{bi} = \int_{-x_p}^{x_n} \vec{E}(x)dx \tag{2.53}$$

The Poisson equation can be used to analyze the electrostatic potential distribution throughout the p/n-junction. With $W = x_n + x_p$ the width of the depletion layer and therefore the capacitance can be also determined by

$$W_D = \sqrt{\frac{2\varepsilon_s}{q}\frac{N_A + N_D}{N_A N_D}(V_{bi} - V_{bias})} \tag{2.54}$$

and

$$C_D = \frac{\varepsilon_s A}{W_D} = \frac{A}{2}\sqrt{\frac{N_A N_D}{N_A + N_D} \cdot \frac{2q\varepsilon_s}{V_{bi} - V_{bias}}} \tag{2.55}$$

for the depletion layer width (see equation 2.54 and the junction capacitance (see equation 2.55), respectively.

With applying a bias to the p/n-Junction, the semiconductor is not anymore in thermal equilibrium. For this reason the single Fermi-energy as well as the mass action law is not applicable anymore, so that quasi-Fermi-levels E_{Fn} and E_{Fp} need to be introduced for the n-doped and p-doped regions.

$$n = N_C exp\left(-\frac{E_C - E_{Fn}}{k_B T}\right) \tag{2.56}$$

$$p = N_V exp\left(-\frac{E_{Fp} - E_V}{k_B T}\right) \tag{2.57}$$

In combination with the continuity equations 2.35a and 2.35b, the current density equations 2.32 and 2.33 and $qV = E_{Fn} - E_{Fp}$, the current-voltage characteristics can be described by the Shockley equation:

$$J = J_n + J_p = J_S\left(\exp\left(\frac{qV}{k_B T}\right) - 1\right) \tag{2.58}$$

with

$$J_S = \frac{qD_p p_{n0}}{L_p} + \frac{qD_n n_{p0}}{L_n} \tag{2.59}$$

To compensate simplifications of the Shockley equation, most commonly an ideality factor n is applied. For a more detailed derivation of the equation see [166].

The Schottky Contact

A Schottky contact is described by a metal-semiconductor interface. Analogue to pn-junctions, a space charge region with a barrier results from the built-in potential due to the alignment of the Fermi levels of metal and semiconductor in contact. For an illustration of the band diagram for a p-Schottky contact and an n-Schottky contact see figure 2.11(a) and 2.11(b), respectively. As the work functions are defined for both, metal and semiconductor, as well as the distance of the Fermi level to the vacuum level (see fig. 2.11(a)), the barrier Φ_B and the built-in potential V_{bi} can be obtained from the work functions and the electron affinity χ of the semiconductor ($q\chi = q\Phi_{n,p} - (E_c - E_{Fn,p})$).

$$V_{bi} = \underbrace{\Phi_M - \chi}_{\Phi_B} - \frac{E_c - E_{F,n}}{q} = \Phi_M - \Phi_n \tag{2.60}$$

for n-doped semiconductors and

$$V_{bi} = \underbrace{\chi + \frac{E_c - E_{F,n}}{q}}_{\Phi_p} - \Phi_M = \Phi_p - \Phi_M \tag{2.61}$$

for p-doped semiconductors with $\Phi_B = \frac{E_g}{q}\chi - \Phi_M$.

For the width of depletion region the derivation is very similar to the pn-junction. However, the densities of acceptors N_A for a p-type and the densities of donors N_D for a n-type Schottky contact can be neglected in equation 2.52 since a metal-semiconductor junction is considered. The depletion width for a p-type ($N_B = N_A$) and a n-type Schottky contact ($N_B = N_D$) is then given by

$$W = \sqrt{\frac{2\varepsilon_s}{qN_B}\left(V_{bi} - V_{bias}\right)} \tag{2.62}$$

With the simple capacitance relation of $C = \varepsilon_s A/W$ equation 2.62 can be modified as

$$\frac{1}{C^2} = \frac{2\left(V_{bi} - V_{bias}\right)}{qN_B\varepsilon_s A^2} \tag{2.63}$$

so that the density of charge carriers can be graphically determined by the slope and the built-in voltage V_{bi} at the intersection with the voltage axes when plotting $1/C^2$ versus V_{bias}.

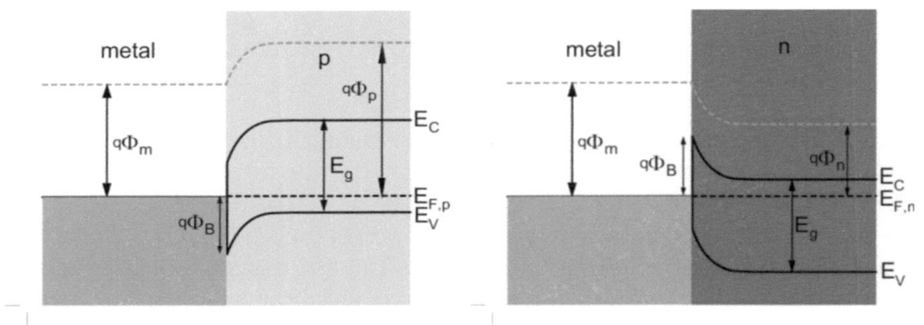

(a) p-type Schottky contact (b) n-type Schottky contact

Figure 2.11.: The Schottky contact for (**a**) a p-doped semiconductor and (**b**) for an n-doped semiconductor. From the energy band diagram, the forming of the barrier and built-in potentials can be derived.

A profound description of the theory can be found in ref. [166, 53]

The Heterojunction

Heterojunctions are of importance for many applications of semiconductors as for example lasers, photodiodes and solar cells. OPDs commonly consist of an organic bulk heterojunction (oBHJ), so that heterojunctions play a major role for organic optoelectronic devices. Here, the consideration of heterojunctions is kept to a schematic one as heterojunctions in IO-HPDs are on molecular and nanometer scale, while IO-HPDs are further characterized by the amorphous photoactive layer. A small bandgap junction and a step heterojunction are shown in figure 2.12(a) and 2.12(b). A similar energetic interaction for organic and hybrid composites is assumed on the nanometer scale.

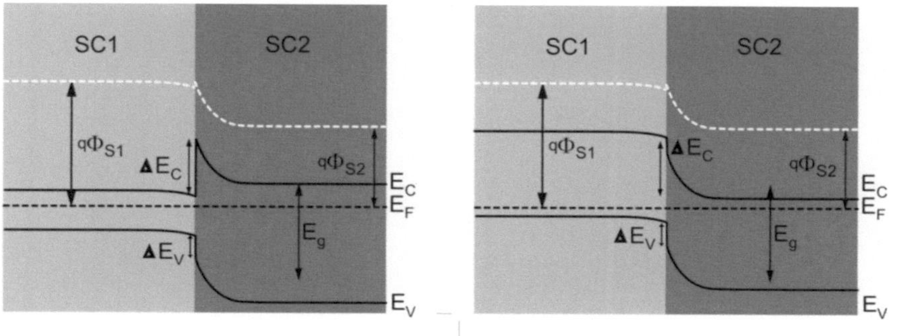

(a) Small Bandgap Junctions (b) Step Heterojunction

Figure 2.12.: Heterojunctions for n-type semiconductors with a smaller bandgap is displayed in (**a**) and for a step heterojunction of two different n-type and p-type semiconductors

Please see again ref. [166, 53] for a deeper explanation.

2.2. Colloidal Quantum Dots of Lead-Sulfide

QDs can be considered as a three dimensional confinement for their charge carriers like a particle in a box (see section 2.1.1), so that instead of a continuous energy band, discrete energy levels are formed within the QDs. The confinement effect and thus the electronic properties strongly depend on the size of the QDs and can be compared to atoms. With *colloidal* QDs it is meant that the QDs are dispersed in a solvent. This is usually achieved by capping the QDs with ligands, so that QDs are homogeneously soluble. Mainly three different types of colloidal QDs can be found in literature: Solely ligand shell capped QDs and type I and type II core-shell colloidal QDs (see figure 2.13). Type I and type II core-shell QDs are characterized in contrast to ligand capped QDs by an additional semiconductor shell surrounding the core. Shells with wider bandgap semiconductors than the core are type I core-shell QDs and shells with semiconductors forming a step heterojunction with the core are type II core-shell QDs. The energy levels in type I and type II core-shell QDs are illustrated in figure 2.14(b) and 2.14(b), respectively [56].

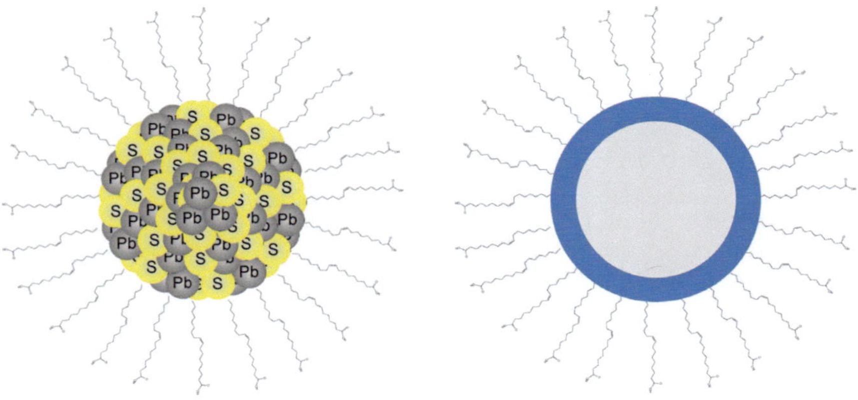

(a) Solely ligand capped QDs

(b) Type I and II core-shell QDs

Figure 2.13.: Different types of colloidal QDs. (**a**) The core of the QDs are capped by ligands only. (**b**) In addition the core is surrounded by another shell of a semiconductor. Type I and Type II differ in the energy profile of the additional semiconductor shell as shown in figure 2.14.

2.2.1. Electro-optical Properties of QDs

QDs are characterized by quantum mechanical confinement, which creates discrete excitonic states above the energy bandgap of the bulk semiconductors. Especially the first excitonic state influences strongly the photonic and electronic properties of QDs. Ref. [13] gave an analytical approximation for the first excitonic state in QDs:

$$E^* \cong E_g + \frac{\hbar^2\pi^2}{2R^2}\left[\frac{1}{m_e^*} + \frac{1}{m_h^*}\right] - \frac{1.8q^2}{\epsilon R} \tag{2.64}$$

with E^* as the energy level of the first excitonic state, R the radius of the QD, m_e^* and m_h^* as the effective masses for electrons and holes, respectively. In figure 2.15(a) the discretization of the energy in QDs is depicted. The confinement introduces discrete energy states above the bandgap which are according to equation 2.64 size-dependent [173]. This confinement can be easily verified

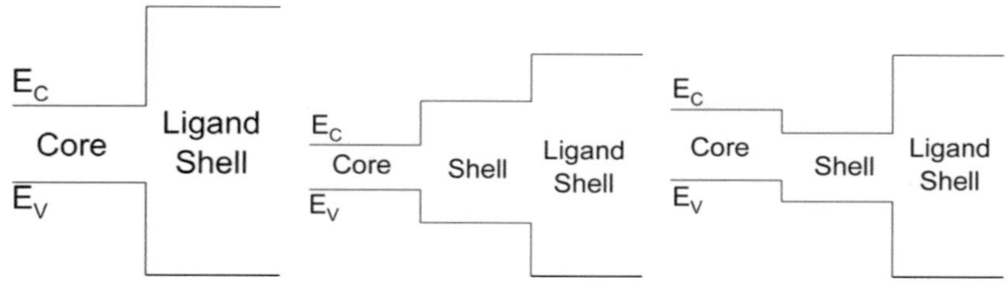

(a) Solely ligand capped QDs

(b) Type I core-shell QDs

(c) Type II core-shell QDs

Figure 2.14.: Energy profile in the different types of QDs. (**a**) For solely ligand capped QDs the core is passivated electronically and confined by the ligand shell only. (**b**) for type I QDs the additional shell is a wide-band gap semiconductor to further passivate and stabilize the QDs. For type II the semiconductor for the shell is chosen to form a step heterojunction between the shell and the core. (**c**) Analogously to the oBHJ a charge separation can be established between the core and shell. In this example the electron is transferred to the shell while the hole remains at the core.

by photoluminescence or absorption measurements. Figure 2.15(b) shows the absorption spectra for different sizes of PbS-QDs clearly indicating the shift of the first excitonic absorption peaks for smaller QDs to shorter wavelengths [55].

Transport and transfer of electrons and holes in QDs are strongly dependent on the passivation of the QDs by their ligand shells. In this work, PbS-QDs with oleic acid (OA) capping are mainly considered, so that these PbS-QDs are highly passivated. For this reason, the coupling of energy states can be only weak as the width of the barrier and therefore the distance to the next molecule or QD is quite large (¿ 1 nm). For a sufficient mobility and charge transport the distance between two QDs should be in the sub-nanometer regime [173]. For PbS-QDs with OA capping weak coupling is assumed only. Under weak coupling conditions Coulomb charging energy has to be considered for charge transfer and transport as well. Ruring charge transport, this energy has to be overcome twice for removing the charge from the first neutral QD to another neutral one. The Coulomb charging energy $E_{Coulomb}$ is given by the following equation:

$$E_{Coulomb} = \frac{q^2}{4\pi\epsilon_m\epsilon_0 R} \tag{2.65}$$

with ϵ_m for the dielectric constant of the surrounding medium. Applying sufficient large bias is a possibility to overcome the Coulomb blockade [173]. Another effect on transport and transfer of charge carriers from QDs is the disorder caused by size distribution during synthesis. In particular, different sizes of

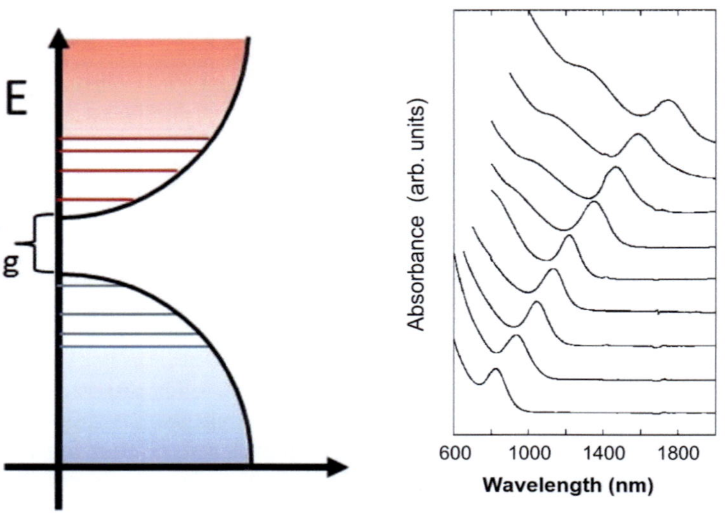

(a) Discretization of States

(b) Absorption Spectra for different PbS-QDs sizes

Figure 2.15.: (**a**) By the confinement of QDs, only discrete energy levels similar to the case of atoms are allowed. The discrete energy levels vary for different sizes of QDs as depicted in (**b**) showing different first excitonic absorption peaks for PbS-QDs. Reproduced from ref. [55].

QDs have slightly different discrete energy states as consequence. However, for resonant tunneling between QDs energetic equal states are necessary. Therefore the energetic matching of materials or a narrow size distribution of QDs from the synthesis is very important for charge transport and transfer.

Since PbS is a small bandgap semiconductor of 0.41 eV, the QDs of PbS are of high interest as the optical properties can be engineered for a wide range of NIR applications. Especially the synthesis of PbS-QDs is already in a state for scaling up to commercial production as describe in the following subsection.

2.2.2. Synthesis

Colloidal Lead-Sulfide (PbS) QDs can be synthesized by various routes. In recent reviews the most promising synthesis was done by the so-called hot-injection method [141, 56, 173, 139]. With this method, a separation of nucleation and growth can be achieved by rapid injection of the reagents into the hot solvent [139]. In figure 2.16 the growth by temporal evolution [31] and by nucleation rate [173] is illustrated. This increases the precursor concentration in the reaction flask above the nucleation threshold. The hot injection leads to an instantaneous nucleation, which is quickly quenched by the fast cooling of the reaction mixture (the solution to be injected is generally at room temperature) and by the decreased supersaturation after the nucleation burst [139]. PbS-QDs received in this work have been synthesized by the hot-injection method according to ref. [55] with OA ligands and showing a narrow size dispersion of 10-15%. Capping with shorter ligands can be done within or subsequently to the synthesis with the goal to improve charge carrier separation from the QDs as well as the enhancement of transport properties is discussed within this thesis.

2.3. Organic Semiconductor Devices

Back in 1977 three pioneering scientists - Alan Heeger, Alan MacDiarmid and Hideki Shirakawa - made an outstanding development. By doping conjugated polymers, the electrical conductivity for polymers (here polyacetylene) was achieved and modified by over 11 orders of magnitude [18]. This was awarded with the Chemistry Nobel Prize "for the discovery and development of conductive organic polymers" in 2000. In 1990 the electroluminescent properties of conjugated polymers were observed by Burroughes et al. [14]. A first solution processed device of a poly(p-phenylene-vinylene) (PPV) layer sandwiched between a low work function (aluminum) and a high work function

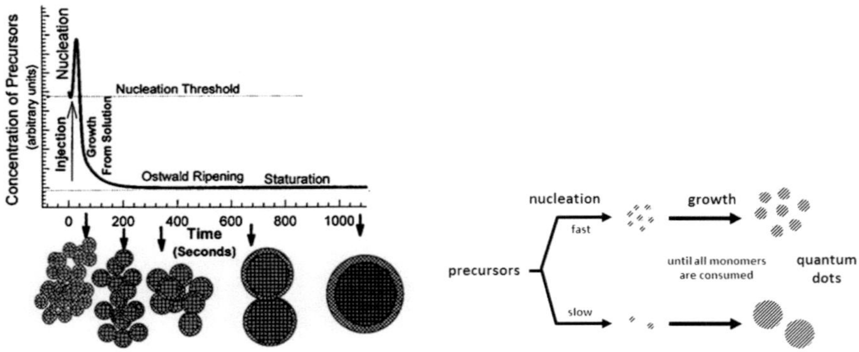

(a) Temporal evolution

(b) Nucleation rate

Figure 2.16.: Controlled growth of QDs by the hot-injection method. (**a**) Ref. [31] and (**b**) ref. [154] give a short insight in the mechanism of growing QDs by hot-injection.

electrode (indium oxide) demonstrated the feasibility of light emitting diodes and a new generation of flat displays. After electroluminescence by conjugated polymers, photovoltaic or photodetection by absorption of organic polymer devices seemed self-evident. However, a different kind of layer, the so-called bulk-heterojunction was developed by ref. [143] to make them possible (see subsection 2.3.2 which was implemented in OPDs by [185]).

All this research, the discoveries and the innovations established a new industry based on new organic semiconductor devices like OLEDs, organic photovoltaics on flexible substrates etc. in the last 20 years, which is still under significant development process.

2.3.1. Conjugated Polymers

Organic Semiconductors can be categorized in two groups: *Small molecules* which are deposited by vacuum sublimation and conjugated polymers. Polymer stems from the Greek words *poly* for 'many' and *meros* for 'part' and implies that it is made of many repeating monomers, so that polymer characteristics strongly depend on the monomers. A *homopolymer* is made of a single monomer whereas a *copolymer* can consist of two or more monomers. *Copolymers* are further discriminated in *alternating copolymers* (figure 2.17 II), *random copolymers* (figure 2.17 III), *periodic copolymers* (figure 2.17 IV), *block copolymers* (figure 2.17 V) and *statistical copolymers*, where the sequences of the different monomers follow statistical rules (e.g. Markovian statistics). In figure 2.17 the

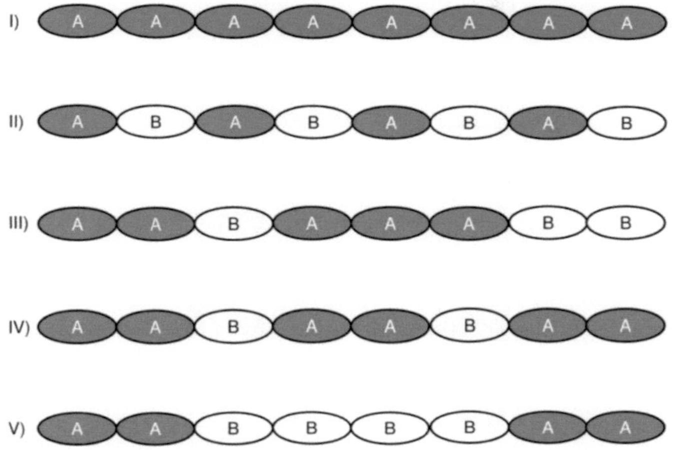

Figure 2.17.: Type of polymers (**I**) Homopolymer: single monomer only. (**II**) Alternating copolymer: two alternating monomers A and B. (**III**) Random copolymer: stochastically ordered monomers A and B. (**IV**) Periodic copolymer: periodically repeating structures of monomers (e.g. A-A-B)$_n$ monomers. (**V**) Block copolymer: covalently bonded homopolymers.

different types of polymers are illustrated. Materials are called organic if they are predominantly consisting of carbon and hydrogen atoms.

Hybridization of atomic orbitals in carbon

The hybridization of atomic orbitals is a special property of carbon atoms, which is essential for organic semiconductors. Figure 2.18 shows the 3 different hybridization states: sp^3-, sp^2- and sp-hybridization. The s- and p-orbitals combine to new energetically equal atomic orbitals [151].

Since the sp^2-hybridization is a special characteristic for conjugated polymers, sp^2-hybridization and bonding is visualized as an example for ethylene. In figure 2.19(a) in the energy diagram the s- and p-orbitals form three energetically equal sp^2-orbital in a carbon atom which can go into bonding with hydrogen atoms and another sp^2-hybridized carbon atom for ethylene (see figure 2.19(b). The 4 hydrogen atoms as well as the two carbon atoms establish a σ-bond

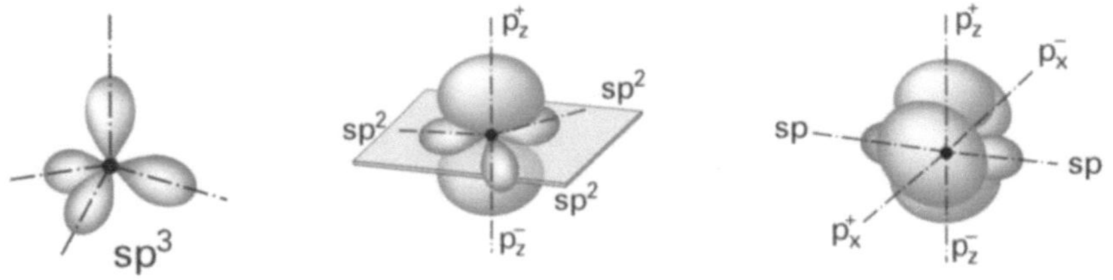

Figure 2.18.: Hybridization form of atomic orbitals for carbon atoms. sp^3- (**left**), sp^2- (**middle**) and sp-hybridization (**right**) is shown. Reproduced from ref. [151].

(rotationally symmetric) [151], which can be seen by the overlap of the corresponding s- and sp^2-orbitals. By the interaction of s- and sp^2-AOs, bonding and anti-bonding MOs are composed: $\sigma(s - sp^2)$, $\sigma(sp^2 - sp^2)$, $\sigma^*(s - sp^2)$ and $\sigma^*(sp^2 - sp^2)$ as indicated in figure 2.19(c). The p-orbitals of the carbon atoms delocalize generating a π bond (not rotationally symmetric) [151]. From this interaction another bonding and anti-bonding MOs $\pi(p - p)$ and $\pi^*(p - p)$ are formed (see figure 2.19(d)). Actually, the carbon are then double bonded. The MOs can be described by a linear combination of AOs, so that for the bonding and anti-bonding MOs the following wave functions are obtained

$$\Psi_\pi = \Psi_{p1} + \Psi_{p2} \tag{2.66}$$

$$\Psi_{\pi^*} = \Psi_{p1} - \Psi_{p2} \tag{2.67}$$

with Ψ_{p1} and Ψ_{p2} being the wave functions of the p-orbitals of the carbon atoms. From figure 2.19 one can anticipate that σ-bonds are stronger compared to π-bonds. σ-bondings exhibit a stronger overlap of the involved orbitals and do not to delocalize as much as π-bonds.

Energy Bands

Conjugated polymers consist not only of a single monomer, but are made by a lot of repeat units as already discussed in figure 2.17. The simplest conjugated polymer can be composed of a unit which is very similar to ethylene. Simply speaking, just by removing an hydrogen from the ethylene molecule another ethylene molecule without an hydrogen molecule can be attached and so on and so forth and a polyacetylene is obtained (c.f. figure 2.20). By increasing the number of repeat units of the polymer the number of bonding and anti-bonding MOs is also increased with the length of the polymer chain. In a

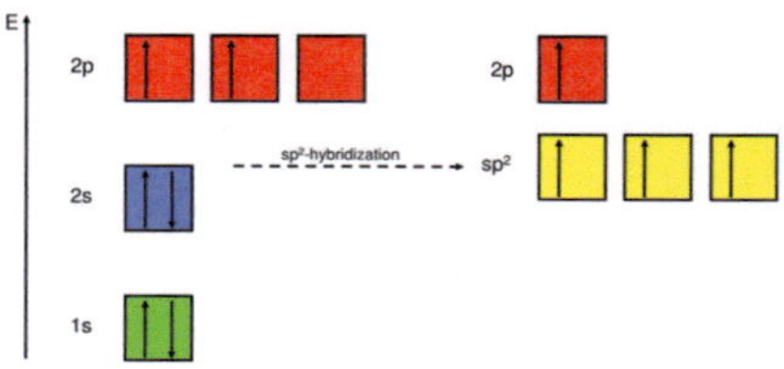

(a) Energy diagram for sp^2-hybridization

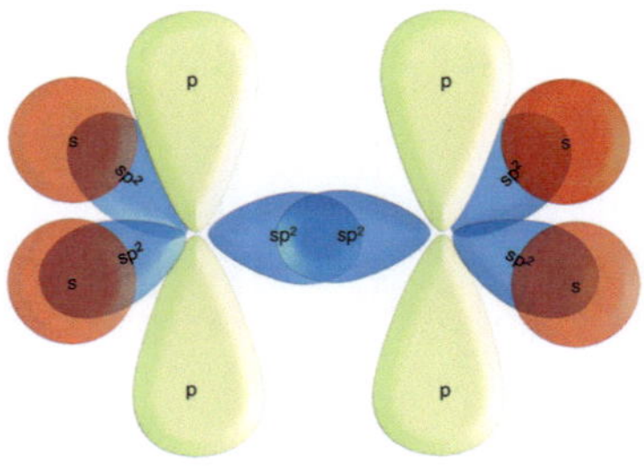

(b) Molecular orbitals in ethylene

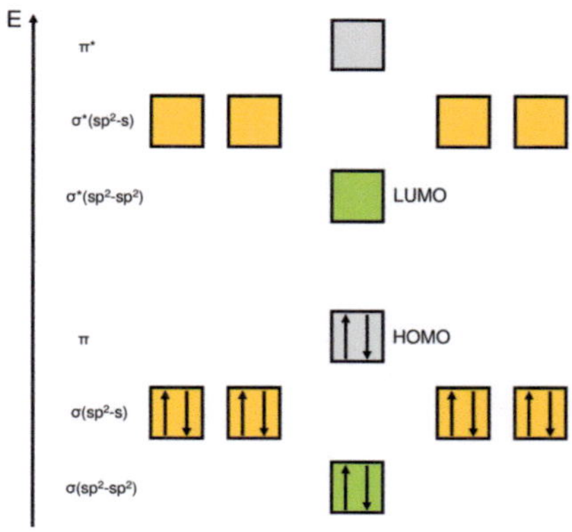

(c) Energy diagram for sp^2-hybridization

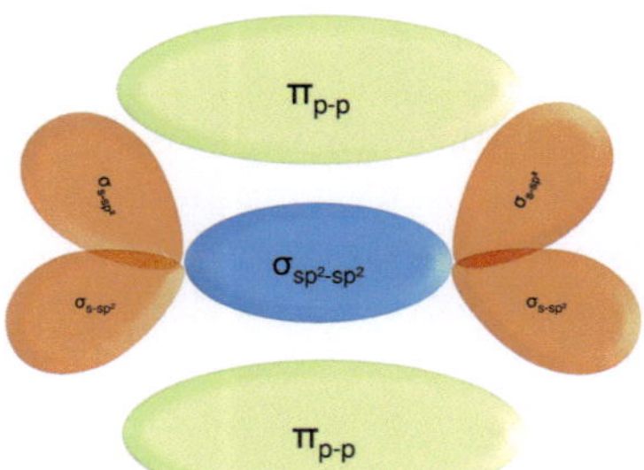

(d) Molecular orbitals in ethylene

Figure 2.19.: Bonding in ethylene molecules with sp^2-hybridized carbon atoms. (**a**) The energy diagram of a sp^2-hybridized carbon atom and (**b**) the involved atomic orbitals (AOs) for bonding in ethylene. (**c**) According to linear combinations of AOs the energy diagram of the molecular orbitals (MOs) can be obtained. (**d**) illustrates the distribution of the MOs in ethylene.

conjugated polymer with N double bonds (π-bonds) 2N linear combination of p-orbitals can be found [104]. At a sufficiently large number of repeat units (carbon atoms) a comparable band configuration to inorganic semiconductors is achieved (see figure 2.21(a)) and a kind of conduction and valence band can be found separated by a bandgap. For organic semiconductor the bandgap is commonly characterized by the difference between the highest occupied molecular orbital (HOMO) and lowest unoccupied molecular orbital (LUMO).

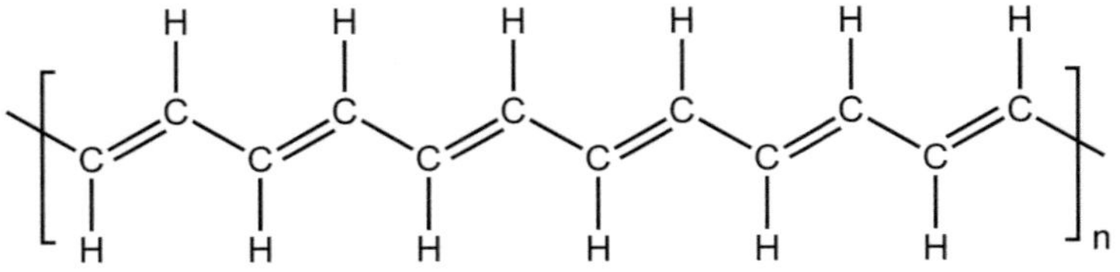

Figure 2.20.: Chemical structure of polyacetylene.

The bandgap is converging at large numbers of carbon atoms and was derived empirically from experimental results according to ref. [120]:

$$E_g = E_0 + \frac{C}{N} \tag{2.68}$$

with E_0 with bandgap for long polymer chains and C as a constant. In 1979, ref. [162] developed a different approach with a tight bonding model concentrating on electron-phonon-interaction α (SSH-model), so that the bandgap in conjugated polymers can be expressed by

$$E_g = 8\alpha u_0 \tag{2.69}$$

with u_0 as the Peierl's distortion. The Peierl's distortion is the disparity in distance between the carbon atoms: $u_0 = d_{single} - d_{double}$ (see figure 2.21(b)). The distance of double bonded carbon atoms is shorter than of single bonded carbon atoms. Without any distortion the bandgap E_g diminishes along with the semiconducting properties of the conjugated polymer.

Conjugated polymers are often mentioned as one-dimensional semiconductors as the charge can travel along the polymer backbone. The electrical conductivity originates from the delocalization of the perpendicular p-orbitals along the chain as illustrated in figure 2.22. Interactions between the p-orbitals smear out the electron probability density throughout the entire polymer chain.

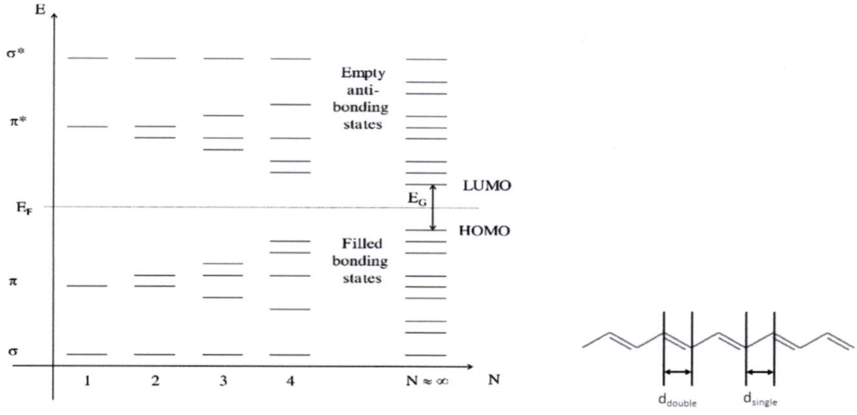

(a) Energy bandgaps in conjugated polymers (b) Peierls distortion

Figure 2.21.: Energy bandgaps in conjugated polymers.(**a**) With increasing the length and thus the number of π-bonds, conjugated polymers form a band structure. (**b**) Peierl's distortion (disparity in distance of the carbon atoms) enables semiconducting properties of a conjugated polymer ($d_{double} < d_{single}$).

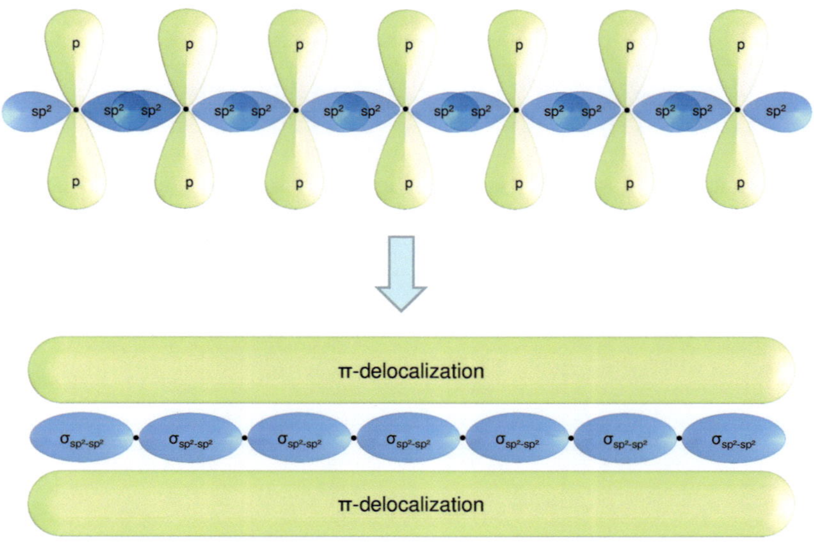

Figure 2.22.: Delocalization of perpendicular p-orbitals along the chain of conjugated polymers (e.g. polyacetylene).

Excited States

Figure 2.23 outlines the formation of a soliton. For polyacetylene two identical mesomeric description of the polymer can be found (see figure 2.23(a)). The A- and B-phase of polyacetylene are only differentiated by the opposite beginning of the alternating sequence of single and double bonds. At the interface of these two phases a soliton at the junction arises. A soliton is an unpaired electron, but electrically neutral. The soliton is free to move as the soliton does not change the system's energy, as the ground state is degenerate. Solitons only occur in degenerate systems. The soliton state is introduced in the middle of the bandgap as depicted in figure 2.23(b). However, semiconducting polymer like polythiophenes show mesomeric forms as benzenoid (c.f. figure 2.24(a)) and quinoid structure (c.f. figure 2.24(b)) and are non-degenerate.

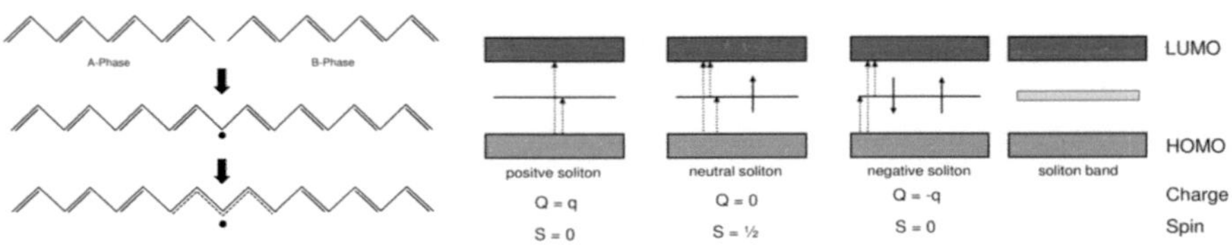

(a) Formation of a soliton (b) Soliton states

Figure 2.23.: (**a**) Formation of solitons and (**b**) introduced states within the bandgap of the conjugated polymer.

For non-degenerate systems polarons can be formed. Polarons are charged and perturb the molecule, so that the energy levels of polarons can be found within in the bandgap as pointed out in figure 2.25. Polarons are usually created by photoexcitation, during charge carrier transport and by chemical doping. If electrons are added or removed from the polymer, negative or positive polarons are generated respectively.

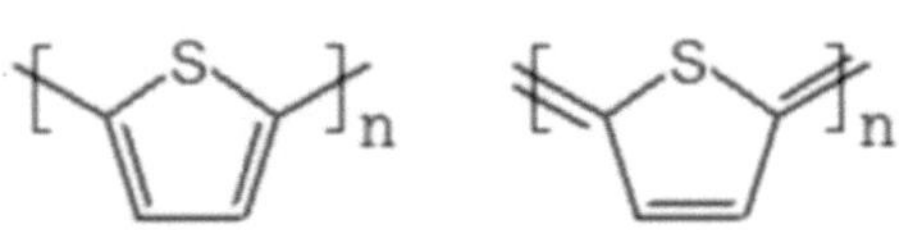

(a) Benzenoid (b) Quinoid

Figure 2.24.: Non-degenerate mesomeric structures of polythiophenes. (**a**) benzenoid and (**b**) quinoid form.

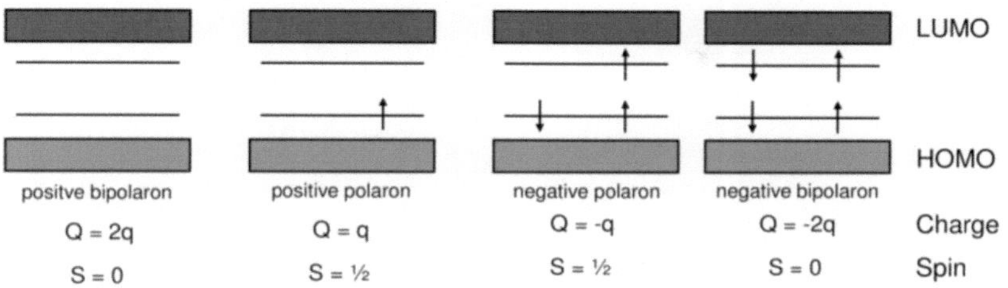

Figure 2.25.: Different types of polarons. Adding or removing electrons to polymers by photoexcitation, charge transport or chemical doping generate positive and negative polarons. Bipolarons can be obtained for sufficiently long polymers by e.g. doping.

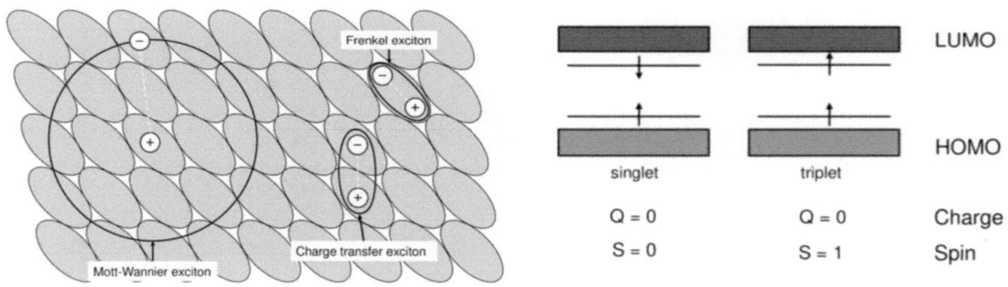

(a) Types of excitons

(b) Singlet and triplet excitons

Figure 2.26.: (**a**) Types and (**b**) states of excitons.

An important mechanism for organic optoelectronics is the photoabsorption in conjugated polymers. An electron is lifted by an incident photon from the π-band to the π^*-band forming an exciton. To generate free carriers the binding energy (the Coulomb interaction) has to be overcome. Excitons can be characterized by three different types: Mott-Wannier, charge transfer (CT) and Frenkel excitons (c.f. figure 2.26(a)). The Frenkel excitons are typically found in molecular systems with a binding radius ¡ 5 Å. Whereas, the Mott-Wannier exciton occur in materials like inorganic semiconductors with large radii of 10 Å to 40 Å [53, 62]. In heterogeneous systems like in oBHJ systems an electron is transferred to a neighboring molecule creating a CT exciton. Furthermore excited states can be characterized by spin-opposition (singlet) or parallelism (triplet). By photoabsorption an electron is raised from a ground state to an excited state. Since the ground state is typically a singlet state, the excited state is also found to be a singlet state. However, by recombination or during charge transport and transfer also triplet states can occur as the spin of those electrons.

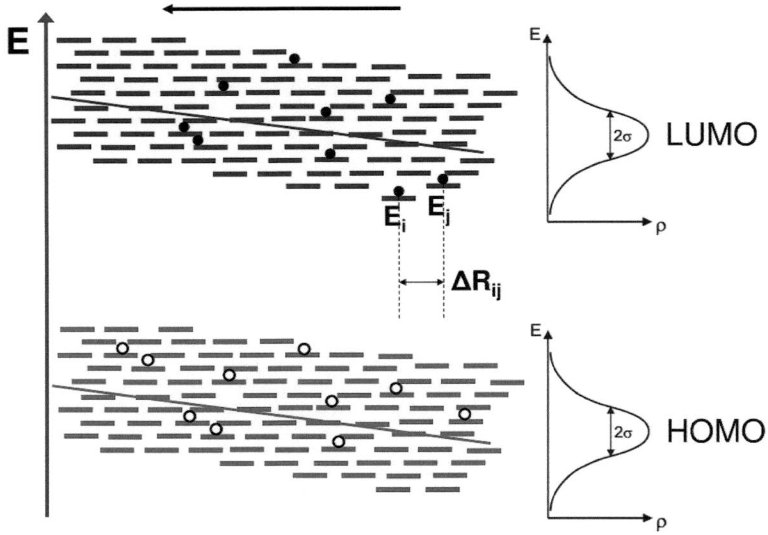

Figure 2.27.: Charge transport in organic semiconductors by hopping along discrete states in the layer.

Charge Carrier Transport

Since organic semiconductors like conjugated polymers usually do not appear in crystalline form, the charge transport is different as compared to inorganic semiconductors. Charge carriers can move freely along the polymer chain only. To travel through a whole layer the charge carriers need to hop to other molecules. Charge transport is therefore dominated by hopping. The states in disordered organic semiconductors are localized and are not continuously delocalized as known from inorganic semiconductors. The carriers have to hop from a localized state in the π or π^*-bands of a molecule to another molecule. This hopping mechanism is a stochastic process. Bssler modeled the charge transport with Monte-Carlo-simulation [7, 8]. Important parameters for the hopping transport consider the Gaussian density of states $\rho(E)$ and the hopping rate ν_{ij} between the states where j is the unoccupied state and i is the donor state, depending on the height of the energetic barrier $E_j - E_i$ and the distance R_{ij} between the states

$$\rho\left(E\right) = \frac{N_t}{\sigma\sqrt{2\pi}} \cdot exp\left(-\frac{E^2}{2\sigma^2}\right) \tag{2.70}$$

N_t represents the total density of states, E is the energy measured relative to the center of the density of states (DOS) and σ is the standard deviation of the DOS. The hopping rate is described by

$$\nu_{ij} = \nu_0 \cdot exp\left(2\gamma R_{ij}\right) \begin{cases} exp\left(-\frac{E_j - E_i}{k_B T}\right) & for E_j > E_i \\ 1 & for E_j < E_i \end{cases} \tag{2.71}$$

where ν_0 is a pre-factor and γ is the inverse localization length of the electron wave function. The temperature and field-dependent mobility can then be described according to ref. [8] as follows

$$\mu_{\sigma,F,\sum} = \mu_0 \cdot exp\left(\left(\frac{2\sigma}{3k_B T}\right)^2\right) \cdot \begin{cases} exp\left(C\left(\left(\frac{\sigma}{k_B T}\right)^2 - \sum^2\right) F^{1/2}\right) & for \sum \geq 1.5 \\ exp\left(C\left(\left(\frac{\sigma}{k_B T}\right)^2 - 2.25\right) F^{1/2}\right) & for \sum < 1.5 \end{cases} \tag{2.72}$$

with a material factor μ_0 and $\sum$ the spatial disorder of the hopping states.

2.3.2. Organic Bulk Heterojunction

The development of the organic bulk heterojunction (oBHJ) was a big step forward for organic photodetector devices, since the concept of oBHJs is the introduction of a heterojunction for charge separation [143]. As in subsection 2.3.1 the different excitons are described by Coulomb bound electron-hole pairs. To dissociate an exciton, more energy than the exciton binding energy is needed. In conjugated polymers Frenkel excitons are generated in an absorption process as they are located on a single molecule characterized by a strong Coulomb binding. Ref. [143] introduced a donor-acceptor-system for a photoactive organic layer. In figure 2.28(a) such a system is formed in this example with the materials Poly(3-hexylthiophene) (P3HT) and [6,6]-Phenyl-C61-butric-acid-methyl-ester (PCBM), which transfers a photoexcited electron from the polymer (donor) to the fullerene (acceptor).

In layers composed of conjugated polymers only, the photoexcitons are Frenkel excitons (see section 2.3.1). Since the electrons and holes are located on a single molecule, recombination is highly probable and lost for extraction. Dissociation of excitons is essential for efficient photodetectors and is realized by an oBHJ. In figure 2.28 an oBHJ is illustrated with PCBM as electron acceptor and P3HT as electron donor. A photoexcited exciton can be separated as the energy difference of the LUMO state of PCBM and P3HT is larger than the binding energy.

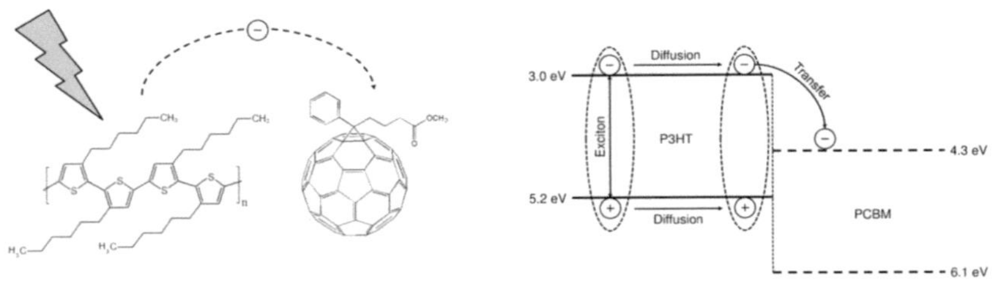

(a) oBHJ (b) Charge Transfer oBHJ

Figure 2.28.: The oBHJ. (**a**) Donor (P3HT) and acceptor (PCBM) molecules can form an oBHJ for charge separation of photoexcitons. Energetic levels and extraction of charges in oBHJs. (**b**) Charge separation can take place at the interfaces, since the energy difference of the LUMOs is large enough to allow charge transfer.

As the exciton diffuses on the polymer (here: P3HT) and as soon as the exciton experiences the energy difference of the acceptor (here:PCBM), the electron is transferred to the PCBM (c.f. figure 2.28(b)) while the hole remains on the P3HT molecule, so that early recombination of the exciton can be avoided. Ref. [143] improved the efficiency of organic solar cells with the introduction of the oBHJ by more than an order of magnitude. The charge separation can be described in more detail by [38]:

I)	$D + A \rightarrow D^* + A$	excitation of the donor
II)	$D^* + A \rightarrow (D - A)^*$	excitation delocalization on D-A complex
III)	$(D - A)^* \rightarrow (D^{\delta+} - A^{\delta-})^*$	initiation of charge transfer
IV)	$(D^{\delta+} - A^{\delta-} \rightarrow (D^{-\bullet} - A^{+\bullet})$	formation of an ion radical pair
V)	$(D^{+\bullet+} - A^{-\bullet}) \rightarrow D^{+\bullet+} - A^{-\bullet}$	charge separation

The mechanism of the oBHJ is applied to a photodiode according to figure 2.29(a). In the photoactive layer of OPDs there are donor and acceptor domains formed during film deposition and annealing by phase separation (see figure 2.29(b)). An exciton is created by an incident photon (c.f. figure 2.29(a) I)) and can be separated as soon as the exciton moved to an interface between donor and acceptor domains (c.f. figure 2.29(b)), the photoexcitons can be separated (see figure 2.29(a) and 2.29(b) II)). Due to the built-in potential because of the low and high work-function electrodes often increased by an additional reverse bias at photodiodes, a field driven extraction of the charge carriers is possible as the hole travels to the anode through the donor phase while the electron is extracted through the acceptor phase at the cathode (c.f. figure 2.29(a) and 2.29(b) III)).

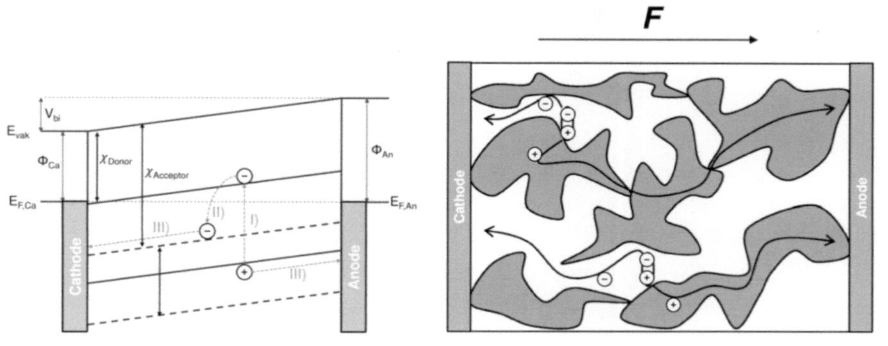

(a) Charge extraction in an oBHJ

(b) Extraction of charges in a BHJ-device

Figure 2.29.: The working principle of an oBHJ in an OPD. (**a**) Energy diagram of an OPD with an oBHJ. The exciton is created by photoexcitation (I), separated at the donor/acceptor interface (II) and extracted at the corresponding electrodes (III). (**b**) In an oBHJ film the morphology is characterized by phase separation into donor and acceptor domains. In the donor domain the exciton is generated, diffuses to a donor/acceptor interface and dissociated (II). The remaining hole is extracted via the donor domains whereas the electron propagates in the acceptor domains to the cathode (III).

2.3.3. Organic Photodetectors

As the important physical principle for photogeneration and efficient charge extraction were explained in the previous sections, device characteristics of OPDs are described in the following.

As already illustrated in figure 2.29(b), an OPD consists of many donor/acceptor interfaces and thus, of many pn-junctions on a molecular level. Such a configuration can be modeled by an "expanded pn-junction" [179]. Accordingly, the I-V characteristics of an OPD can be described by the Shockley equation as already introduced for pn-junctions (c.f. 2.58).

$$I_D = I_S \left(\exp\left(\frac{qV)}{nk_BT} \right) - 1 \right) \tag{2.73}$$

with n as the ideality factor for diodes I_S as the saturation current accounting for thermal assisted injection, and I_D as the diode's current. However, this equation does not consider the additional current by photogeneration of charge carrier.

Taking Beer's law from equation 2.42 into account, all absorbed photons can contribute to the current, so that the maximum (optical possible) current I_{max}

can be extracted. Depending on the wavelength λ and the thickness of the photoactive layer d, I_{max} is given by

$$I_{max}(d,\lambda) = \frac{qP_{opt}A}{h\nu} \cdot (1 - exp\left(-\alpha\left(\lambda\right) \cdot d\right)) \tag{2.74}$$

Losses for oBHJ layers are negligible as all absorbed photons are generating charge carriers (assuming an internal quantum efficiency $\eta_{int} \approx 1$). Due to the finite lifetime τ of photogenerated charge carriers, the charge carriers can recombine on the extraction path. The photocurrent I_{ph} can be therefore described as proposed by [148]:

$$I_{ph} = \begin{cases} -|I_{max}| & \mu\tau(-V + V_{bi})/d > d \\ |I_{max}| & \mu\tau(V - V_{bi})/d > d \\ |I_{max}| \cdot \mu\tau(V - V_{bi})/d^2 & else \end{cases} \tag{2.75}$$

Under large enough forward biases the schubwegs are longer than the film thickness d, so that the all the generated charge carriers can be extracted. For smaller schubwegs, a fraction can be extracted accordingly.

Actually an OPDs as a device consists among others of a glass substrate, electrodes, encapsulation introducing losses. The overall spectral response can be characterized by the external quantum efficiency (EQE) and is a key performance indicator for photodetectors. The EQE puts the extracted photogenerated charge carriers in relation to the incident photons on the detector. OPDs show typically EQEs of more than 75% [147, 176]

$$\eta_{EQE} = \frac{I_{ph}/e}{P_{opt}/hf} \tag{2.76}$$

When the OPD is operated in photovoltaic mode (IV. quadrant of the I-V characteristics), the fill factor (FF) is a standard indicator. In an ideal photodiode FF is 100% as the photodiode could be operated at an power level of $I_{SC} \cdot V_{OC}$. Taking losses into account the current $I_{P,max}$ and the voltage $V_{P,max}$ at the optimum operating power level are smaller than I_{SC} and V_{OC}. The relation is expressed by FF as following:

$$FF = \frac{I_{P,max} \cdot V_{P,max}}{I_{SC} \cdot V_{OC}}$$

Figure 2.30 illustrates the fill factor, the definition of I_{SC} and V_{OC} and the photocurrent I_{ph}. From the I-V characteristics an equivalent circuit can be derived from parameters (e.g. R_p, R_s) as depicted in figure 2.31. Material and device properties as leakage current due to imperfectly blocking contacts and bulk

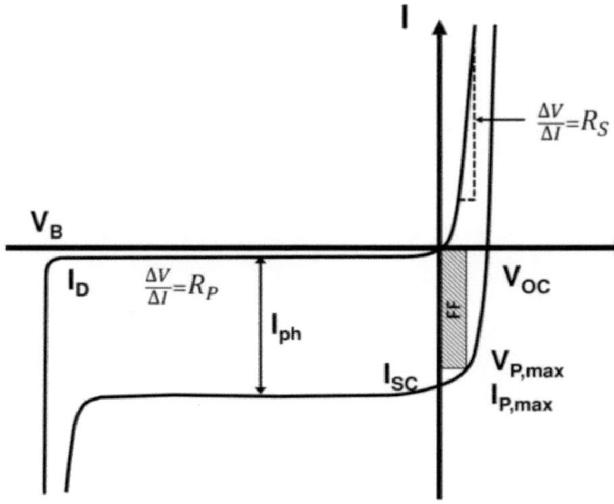

Figure 2.30.: Typical I-V characteristics of an OPD.

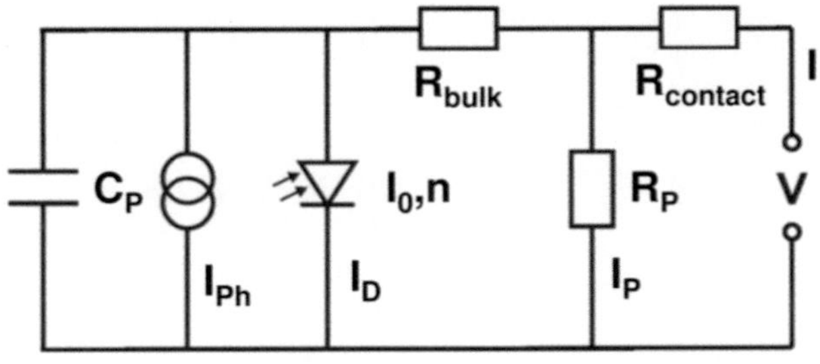

Figure 2.31.: Equivalent circuit of an OPD.

resistance and contact barriers can be taken into account for I-V characteristics.

Consequently, the I-V characteristics of an OPD can be implicitly, analytically described from the equivalent circuit by

$$I = I_S \left(\exp\left(\frac{q(V - IR_S)}{nk_B T} \right) - 1 \right) + \frac{V - IR_S}{R_P} + I_{ph} \tag{2.77}$$

P3HT:PCBM devices were modeled by ref. [148] showing a good agreement with measurements

2.4. Hybrid Organic/Inorganic Photodetectors with Colloidal QDs

QDs and colloidal quantum dots (CQDs) were introduced in section 2.2 and can be applied to optoelectronic devices. For photodetection two main approaches can be considered: Devices with photoactive layers containing purely CQDs and with hybrid composites [88].

Hybrid devices

In hybrid devices, the film is not a QD solid, but a composite of usually semiconducting organic polymers and CQDs. To obtain hybrid composites for processing different methods have been found in research.

Blending the semiconducting polymers with the QDs together and process layers from the solution, is an obvious step. However, limitations of this approach have to be taken into account. The materials have to be carefully chosen according to solubility, phase separation of the composites in the films, which is crucial for the morphology, and the essential charge carrier separation and transportation within the films. For these reasons properties of the QDs and of the polymer have to be adjusted (e.g. for solvents, energy level alignment and charge transfer). In this work, this approach was applied to develop the IO-HPD.

A second approach is to link polymers directly to the QD. Before solution processing of hybrid films, the initial ligand of the QD is exchanged with functionalized polymers. The polymers have side or end anchoring groups to replace the initial ligands and passivate the surface of the QD. To skip the process for ligand exchange, the anchoring groups can be integrated to the ligand.

The most integrated solution to generate hybrid composites is to use already a solution of polymers as growth media for the CQD synthesis: a so-called in-situ growth of QD. The polymers passivate the QD and films can be processed from the solution. This approach was demonstrated for several materials as TiO, ZnO, PbS and CdS. Power conversion efficiencies in solar cells of 2.9% have been reached [139]. For more detailed information about creating chemically hybrids see also ref. [139].

Selected IO-HPDs reported in literature

Ref. [105] realized an hybrid photodetector with PbS-QDs embedded into a polymer matrix of poly[2-methoxy-t-(2'-ethylhexyloxy)-p-phenylene vinylene (MEH-PPV). To facilitate charge transport on the QDs the initial OA ligands were exchanged with octylamine ligands. EQEs of 0.38% at 975 nm were reported for this system. Here, the QDs are utilized as electron acceptor, since MEH-PPV is an hole-conducting polymer used in organic photovoltaics [58].

To sensitize PCBM films with PbS-QDs was shown by ref. [172]. The PbS-QDs were fully capped with OA ligands, but showing ultra-fast charge transfer from QDs to the fullerene. The phototransistor device of ref. [172] showed a significant higher responsivity and improvement of the photoresponse to ref. [105].

Ref. [16] incorporated CdTe-QDs in an oBHJ matrix of P3HT:PCBM showing a high photoconductive gain and efficiencies of 8000% and 600% at 350 nm and 700 nm respectively. A similar approach for the visible range coincided with the work for IO-HPDs sensitized with PbS-QDs for the near-infrared.

A further bulk-heterojunction considering the principle of the Graetzel photovoltaic cell was recently reported by the Sargent group in collaboration with Graetzel reaching power conversion efficiencies of 5.5%. PbS-QDs are the sensitizers for the porous titanium oxide layer.

CQD only devices

Photodetectors just made of QD layers were investigated by several groups. In 2000, ref. [93] reported photoconductivity of differently passivated CdSe-QDs on a thermally passivated silicon substrates with Au electrodes in lateral configuration. A few years later, ref. [119] demonstrated EQEs of 24% for a sandwiched CdSe-QDs in indium tin oxide (ITO):Poly(3,4-ethylenedioxythiophene) (PEDOT):Poly(styrenesulfonate) (PSS):CdSe-QD:Silver (Ag) structure. In fact, these devices were still photoconductors without any photodiode properties. Ref. [10, 9] have shown long-wavelength operation for photodetection with inkjet-printed dodecanethiol capped HgTe-QDs on interdigitated Ti/Au electrodes with an alternative processing method of inkjet printing. After first studies of hybrid devices [105], the group of Sargent have done distinctive work in the recent years on single QDs devices with a focus on lead salt QDs. Ref. [86] accomplished with a photoconductor of PbS-QDs in a lateral configuration a detectivity (D*) exceeding epitaxially grown InGaAs detectors of 10^{12} jones. Ref. [85] reported a photovoltaic octylamine and dithiol capped PbSe-QD device with an EQE of 46% in the infrared. However, the lifetimes of these device were in the range of several days, only. An optimized device with PbS-QD for photodection in a sandwich configuration (ITO/PbS-QD/Al) was developed by ref. [26]. The PbS-QDs were surface passivated by a 3-stage procedure with benzenedithiol. Characteristics of a photodiode were achieved with low dark currents ($0.1\,\mathrm{nA\,cm^{-2}}$), infrared sensitivity up to 1600 nm (EQE $> 10\%$) and therefore an D* of $> 10^{12}$. Fast operation with a -3 dB frequency of more than 10^6 Hz could be demonstrated as well. Although lifetime was improved to more than two months, for applications the lifetime has still to be dramatically improved.

Detailed review on photodetectors made of colloidal quantum dots are given for further readings from ref. [139, 141, 56] and ref. [142, 174, 88].

3. Measurement and Characterization Methods

In this chapter an overview on the used characterization and measurement methods is given. The chapter is divided into three parts describing methods to investigate device, structural and material / blend properties.

3.1. Device Properties

Within this work devices are analyzed by means of electro-optical characterization and transient measurements. The stability of devices is determined by accelerated lifetime measurements. As the photodiodes are integrated as well into imaging systems, capturing methods and imaging properties are discussed elements in this section.

3.1.1. Electro-optical Characterization of Device Properties

Essential figure of merits are electro-optical properties of photodiodes. Photodiodes are analyzed in terms of current-voltage (I-V) characteristics, external quantum efficiencies and spatially resolved photocurrents within this work, too.

Current Voltage Characteristics

In this measurement the currents of the device under test (DUT) are analyzed in dependence of the applied voltage.

For a standardized layout of testdiodes (see 3.2) containing 8 devices the first setup in figure 3.1 is applied. A switch (Keithley 7001) addresses each individual diode and an Ampere meter (Keithley 6487) acquires the I-V characteristics via a LabVIEW interface with a resolution of ~ 100 pA. After acquiring the

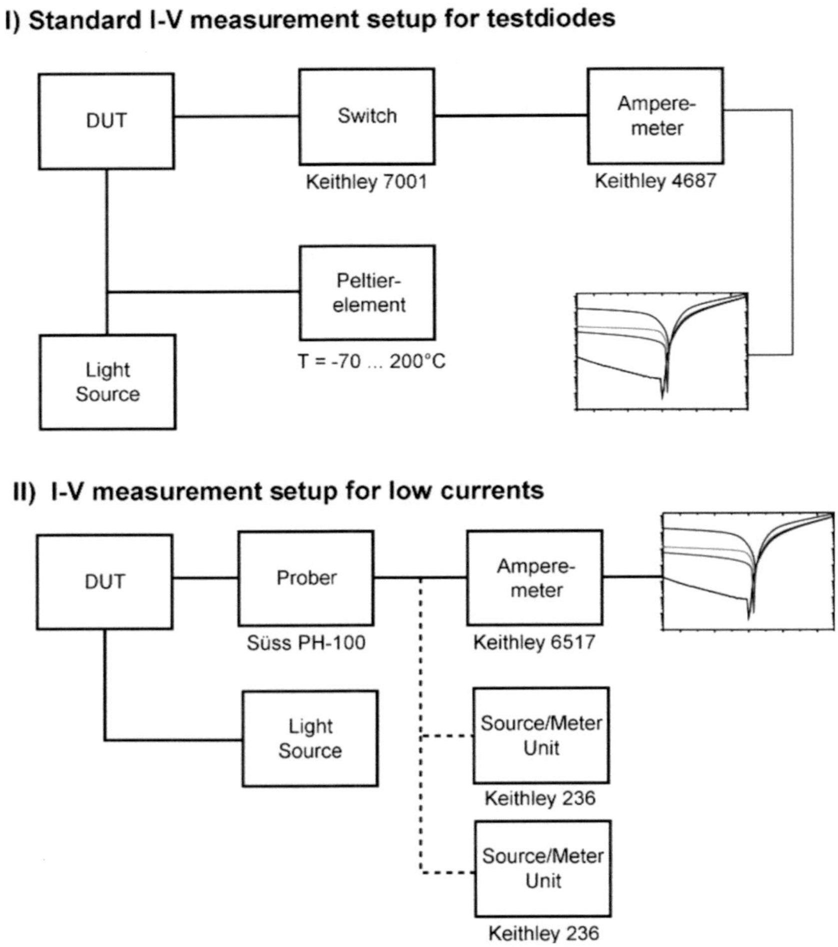

Figure 3.1.: Setups for measurements of I-V characteristics. (**I**) A standardized semi-automated measurement setup is used for standard testdiodes for I-V characterization in the dark and with illumination (see table 3.1). (**II**) For customized and for high-resolution I-V measurement a second setup is applied accordingly.

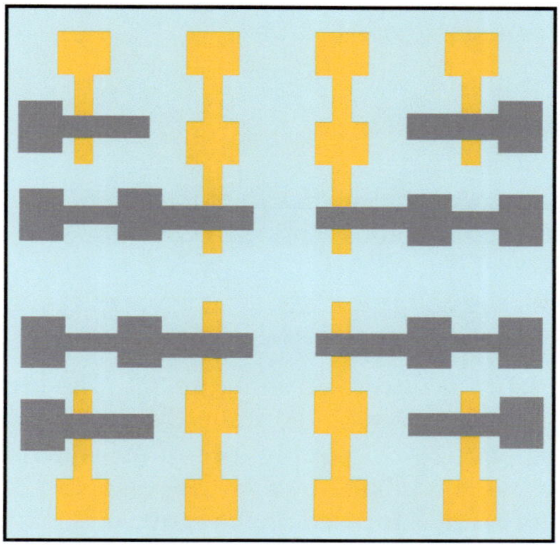

Figure 3.2.: A standard layout for 8 testdiodes of 2 cm × 2 cm.

dark currents, the samples are illuminated subsequently to determine photoresponses. Types of light sources for illumination are given in table 3.1. If not otherwise stated, I-V measurements are recorded by the above presented procedure. Optional, a Peltier element can be included to conduct temperature dependent measurements ranging from −70 °C to 200 °C.

To measure non-standard devices, a second setup with probe tips (Sss Microtec PH 100) is available. This setup is also used to measure low currents down to a few hundreds fA with a Keithley 6517 Ampere meter. To measure more complex circuits the setup is extended with two Keithley 236 source meter units. A possibility to couple light into the DUT by an optical waveguide is also ar-

Light source	Spectra	Wavelength (nm)	Intensity (µW cm^{-2})
Green LED array	Monochromatic	532	780
NIR LED array	Monochromatic	880	∼500
Halogen lamp	Polychromatic	Solar Spectrum AM 1.5G	Peak: 1100
Filtered halogen lamp	Polychromatic	> 870	n.a.

Table 3.1.: Different types of illumination sources applied to measurements of I-V characteristics.

ranged. Measurements with this setup are specifically mentioned throughout this work for transparency reasons.

External Quantum Efficiencies

A significant performance indicator for photodetectors is the ratio of collected charges N_{col} and incident photons N_{Ph}, the EQE.

$$EQE_M(\lambda) = \frac{N_{col}}{N_{Ph}} = \frac{I_M}{I_R} \cdot EQE_R(\lambda) \tag{3.1}$$

The measurement setup is illustrated in figure 3.3. The broad spectrum light source is a halogen lamp which is spectrally dispersed by a monochromator (LOT Oriel Cornerstone 130). To apply lock-in measurement techniques the monochromatic light is modulated by a chopper (HMS 220). The modulated monochromatic light beam is focused on the active region of the OPD by a lens and a mirror. The photogenerated current is AC amplified, voltage converted (Femto DHPCA-100) and acquired by a lock-in amplifier (EG&dG 5210). The measured current I_M is put into relation of the previously measured reference current of the calibrated device I_R (see equation 3.1). The calibrated photodiode is a silicon photodiode for wavelengths up to 1050 nm and an InGaAs photodiode for the visible-NIR range from 650 nm to 2500 nm.

Measurement of X-ray Response by Lock-in Technique

In this work the sensitivity of hybrid devices to X-ray exposure is investigated as well. To measure the response of the device an analogue procedure to the EQE-measurements by a lock-in technique is applied. The monochromator and the light source are replaced by an x-ray tube operated at 50 kV as acceleration voltage. To provide a chopped signal for lock-in measurements the chopper wheel was enforced by additional lead metal sheet of 250 µm to ensure sufficient x-ray absorption. Note that the x-ray tube, the chopper wheel and the hybrid device were located in an x-ray sealed chamber for protection while the measurement equipment was kept outside.

Spatially Resolved Photocurrents

In figure 3.4 the schematic of the measurement setup for a spatial resolution of the photocurrents is depicted. For optical excitation of the DUT two different light emitting diodes (LEDs) emitting at 532 nm and 940 nm, respectively are applied and coupled into a microscope to focus the light on the active area

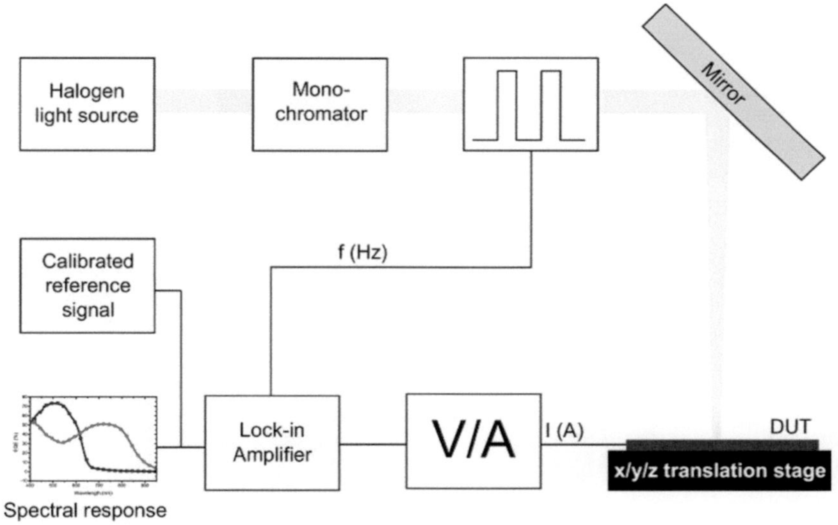

Figure 3.3.: Diagram of a setup for spectral EQE measurements. The incident light from a halogen lamp passes through a monochromator and is modulated by a chopper wheel. With a mirror and a three-dimensional translation stage the DUT is placed accurately in the beam. The generated modulated photocurrent is amplified by a current-voltage amplifier and measured with a lock-in amplifier. By setting the measured and the reference signal into relation, the EQE of the DUT can be determined.

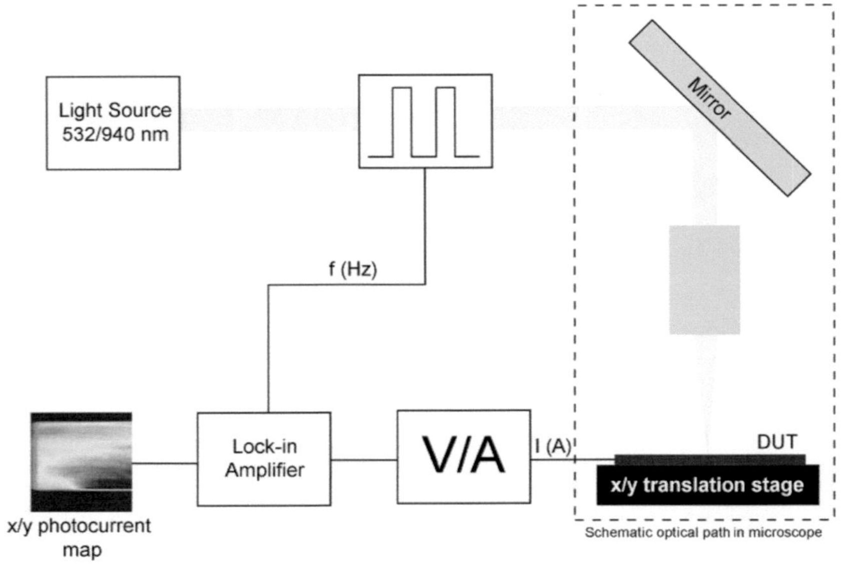

Figure 3.4.: Setup for spatially resolved photocurrents. Two different LEDs at 532 nm and 940 nm are used as light source and is focused on the active area of the DUT as small as a 3 µm spot. The optical signal is modulated to use lock-in technique. Scanning of the DUT is done by the x/y translation stage.

of the photodetector. The focused spot is reduced to a size of about 3 µm or 10 µm in diameter for the 532 nm emitting LED and for the 940 nm emitting LED, respectively. Since small photocurrents have to be measured the lock-in technique is used by modulating the optical signal at a frequency of 70 Hz. The photocurrents are amplified and converted by a current-voltage amplifier. The resolution in x- and y-direction for scanning the active area of the photodetector is as small as 5 nm.

3.1.2. Transient Measurements

Pulsed Measurements

Within this work two different optical pulsed measurements are used. Figure 3.5 illustrates a setup for pulsed measurements applied to standard test diodes (see layout 3.2) while the setup of figure 3.6 for specially designed diodes for high frequency measurements (see layout in figure 3.7).

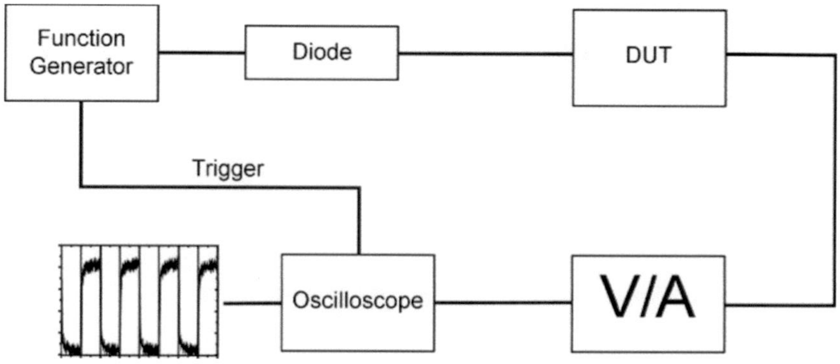

Figure 3.5.: Setup for optical pulsed measurements

For pulsed measurements of standard testdiodes the pulse is applied to an LED (Lumileds Luxeon 3 Star) emitting at 532 nm by a function generator (TTi TGA1242). The resulting currents of the photodiodes are amplified and converted by Femto DHPCA-100, subsequently acquired and displayed by an oscilloscope (LeCroy Wavesurfer 424) as shown in figure 3.5.

For high frequency measurements the diodes do not have any contact wires to the active area and are contacted by an high frequency probe as used by ref. [130, 129]. As depicted in figure 3.6 a light pulse of an Nd:YAG-Laser emitting with 2.5 ns at 532 nm follows an optical path through an neutral density filter to apply different optical powers and a beam splitter to direct the light on to the DUT and to an optical power meter for monitoring the optical power. For I-V measurements and applying bias a source-meter-unit (Keithley SMU 236) is used. The signal traces are acquired by an oscilloscope (Agilent Infiniuum 54832B) with a maximum sampling rate of 4 GS/s (0.25 ns) triggered by a fast photodiode.

From transient measurements parameters as the cut-off frequency f_{-3dB} (signal decreased to factor of -3 dB), the rise time t_r and the fall time t_f (signal raises from 10% to 90% of the steady state values or declines vice versa) can be derived.

Noise Analysis

Noise is a very important detector characteristic as noise limits the resolution or the sensitivity of a detector, respectively. Noise is the result of random stochastic processes, so that the perturbation of the current's amplitude by noise is totally uncorrelated. Noise is defined as the standard deviation of the variable

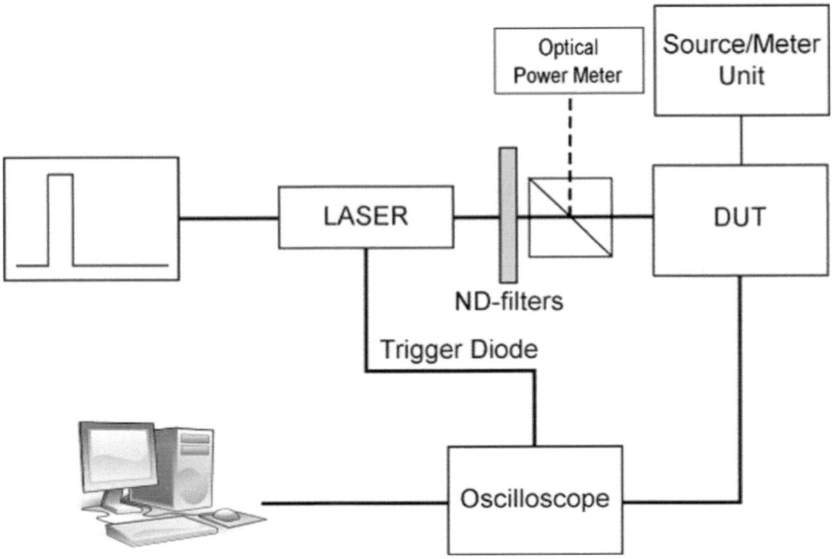

Figure 3.6.: Setup for high frequency optical pulsed measurements

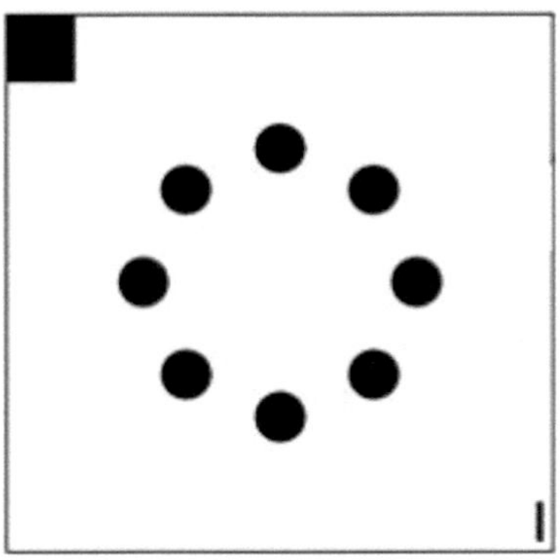

Figure 3.7.: Layout of testdiodes used for fast transient measurements reproduced from ref. [129].

(here: the current) and the variance (standard deviation squared) is described by

$$\overline{(\Delta i)^2} = \overline{(i - \hat{i})^2} = \frac{1}{\Delta t}\int_0^{\Delta t}\left(i - \hat{i}\right)^2 dt \tag{3.2}$$

with Δt as a time interval long enough for any noise signals and $\hat{i}$ as the average current.

Another property is that each independent source of noise sums up by square to the total noise current, since their variances and accordingly the noise power spectra and not the noise currents are summed up. The total noise power for N independent sources is then given by

$$\overline{(\Delta i)^2} = \sum_{i=1}^{N}\overline{(\Delta i)_i^2} \tag{3.3}$$

the root-mean-square-value of the noise current Δi_{rms} is

$$\Delta i_{rms} = \sqrt{\overline{(\Delta i)^2}} = \sqrt{\sum_{i=1}^{N}\overline{(\Delta i)_i^2}} \tag{3.4}$$

If the noise is uncorrelated to the frequency (white noise), the noise power is proportionally dependant on the bandwidth Δf and

$$\Delta i_{rms} \propto \sqrt{\Delta f} \tag{3.5}$$

The main sources for noise in photodetectors are the following according to ref. [165, 166]:

- **Johnson noise** is a thermally induced random motion of charges in any linear resistor. Thermal noise is independent of frequency and has a constant noise power density throughout the frequency spectrum.

$$I_J = \sqrt{4k_B T R \Delta f} \tag{3.6}$$

- **Shot noise** originates from a discontinuous current flow in the detector since charge carriers promote only a discrete charge and generate time-dependent fluctuations in the electrical current. The fluctuations are independent from frequency (white noise spectrum) for low and intermediate frequencies.

$$I_S = \sqrt{2q\hat{i}\Delta f} \tag{3.7}$$

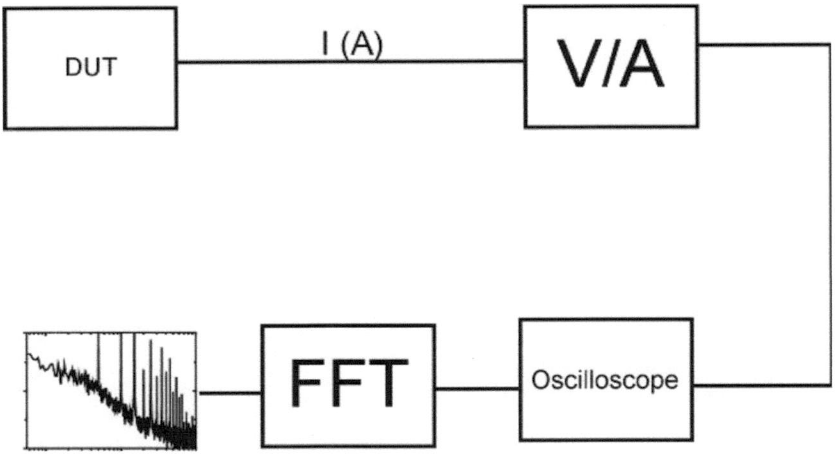

Figure 3.8.: Setup for noise analysis. With a current-voltage amplifier (Stanford Research Systems, SR570) the acquired time-resolved signals by the oscilloscope (LeCroy Wavesurfer 424) are transformed by fast Fourier transformation (FFT).

- **Flicker noise (1/f noise)** is inversely proportional to the frequency and is therefore not considered as white noise. Processing or material properties of the device are influencing flicker noise as it is related to impurities or surface states. At frequencies below 10 kHz flicker noise becomes important and may dominate other noise sources.

$$I_F = \sqrt{\frac{K I_{dark} \Delta f}{f}} \tag{3.8}$$

  K is an empirically determined constant which differs largely and is dependent on material and device geometry.

For an OPD the noise current is then given by

$$I_{noise,PD} = \sqrt{I_J^2 + I_S^2 + I_F^2} \text{ in } \left(\mathrm{A}\sqrt{\mathrm{Hz}}\right) \tag{3.9}$$

Two major performance indicators to compare photodetectors are the noise equivalent power (NEP) and the specific detectivity D^*. The NEP is defined as the incident optical power to generate a photocurrent in the photodetector which is equal to the noise current $I_{noise,OPD}$. The NEP is described by

$$NEP = \frac{I_{noise,PD}}{R_{PD}(\lambda)} \text{ in } \left(\mathrm{W}\sqrt{\mathrm{Hz}}\right) \tag{3.10}$$

with R_{PD} as the photodetectors spectral responsivity. The spectral responsivity of a photodetector is given by

$$R_{PD}(\lambda) = EQE(\lambda) \cdot \frac{q\lambda}{hc} \tag{3.11}$$

resulting with equation 3.10 in

$$NEP = \frac{I_{noise,PD}}{EQE(\lambda)} \cdot \frac{hc}{q\lambda} \approx \frac{I_{noise,PD}}{EQE(\lambda)} \cdot \frac{1240\ \frac{\mathrm{W \cdot nm}}{\mathrm{A}}}{\lambda(\mathrm{nm})} \tag{3.12}$$

Since noise currents are dependent on different parameters, the NEP is only helpful, when these parameters are accounted for. The frequency, the temperature and the dark currents (applied bias) during the measurement of the photodetector strongly influence noise currents. Whereas, the responsivity of the photodetector depends on the wavelengths of the incident light. The detectivity D is the reciprocal of the NEP, so that a greater detectivity expresses a higher resolution of photodetectors for lower optical powers. Since this consideration is independent from the active area of the photodetector, the specific detectivity D^* is introduced.

$$D^* = \frac{\sqrt{A_{PD}}}{NEP} = \frac{R_{PD}\sqrt{A_{PD}}}{I_{noise,PD}} \text{ in } \left(\mathrm{cm}\sqrt{\mathrm{Hz}}\ \mathrm{W}\right) \tag{3.13}$$

By an FFT of an oscilloscope (Stanford Research Systems, SR570) the acquired noise currents of the photodetector are determined in frequency domain. In combination with EQE-measurements (see section 3.1.1) the NEP and the specific detectivity D^* can also be derived for photodetectors from these measurements.

Impedance Spectroscopy

Impedance spectroscopy is a method to analyze the impedance $Z(2\pi f)$ of a device with relation to the frequency domain. The impedance is characterized by the relation of an alternating voltage and its corresponding current response or vice versa. This can be expressed by

$$Z(2\pi f) = \frac{V_0 \cdot \sin(2\pi f)}{I_0 \cdot \sin(2\pi f + \phi)} = Z_0(2\pi f)\exp(j\phi 2\pi f) \tag{3.14}$$

$$\tag{3.15}$$

so that the impedance $Z(2\pi f)$ can also be split in a real and an imaginary part as known from complex mathematics. Actually, the impedance of an electrical

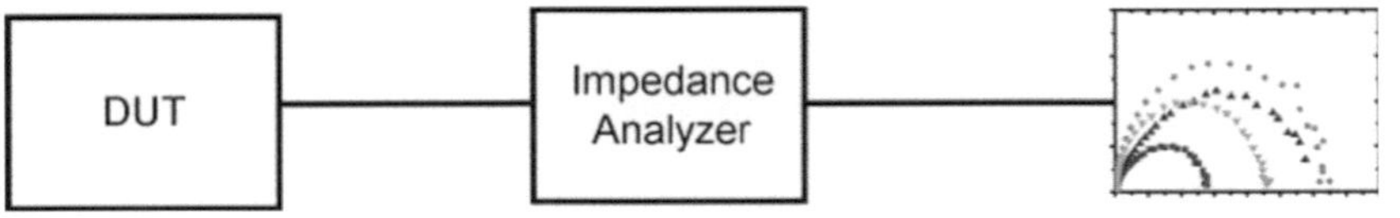

Figure 3.9.: Setup for impedance spectroscopy measurements. The impedance analyzer is the model HP 4192A from Hewlett-Packard is used.

circuit can be analytically determined from serial and parallel combinations of resistances, capacitances and inductances.

$$Z_R = R \qquad \text{Resistance} \tag{3.16}$$

$$Z_C = \frac{1}{j \cdot 2\pi f \cdot C} \qquad \text{Capacitance} \tag{3.17}$$

$$Z_L = j \cdot 2\pi f \cdot L \qquad \text{Inductance} \tag{3.18}$$

From the spectra of the impedance of the device, conclusions on an equivalent circuit can be made for small signals as during impedance spectroscopy the oscillating amplitude of the frequency dependent signal is about 100 mV. For interpretation and analysis of the frequency spectra see section 7.3.2.

3.1.3. Imaging

Since the goal of this work is to develop an NIR imager, the capturing from an imaging device is necessary as well as methods to evaluate imaging performances.

Figure 3.10 depicts the scheme of a setup to capture images from an a-Si AM-TFT panel. The readout integrated circuit (ROIC) consists of the addressing and timer logic to drive the a-Si AM-TFT panel, charge amplifiers and a 14-bit analog digital converter (ADC). Since the electronics is highly integrated the signals are simultaneously integrated, amplified, low-pass-filtered and A/D converted. A more detailed view of the ROIC and its hardware implementation is shown in figure 3.11. The entire electronics is able to drive and read 256 lines. The charge amplifiers are adjustable in their amplification by selectable integrating capacitances of 0.5 pF, 1 pF, 2 pF and 4 pF. Actually, for the NIR imagers a value of 4 pF had to be chosen due to the high dark currents. With this setting 1 count of the 14-bit ADC corresponds to 4083 collected electrons. The integration time for each pixel row is fixed to 20 µs. The digital signal can be transferred on a computer with a grabber card by a camera link interface for image processing. To perform NIR imaging, an InGaAsP laser diode (ML725B8F) as an external illumination source is used. The simple method

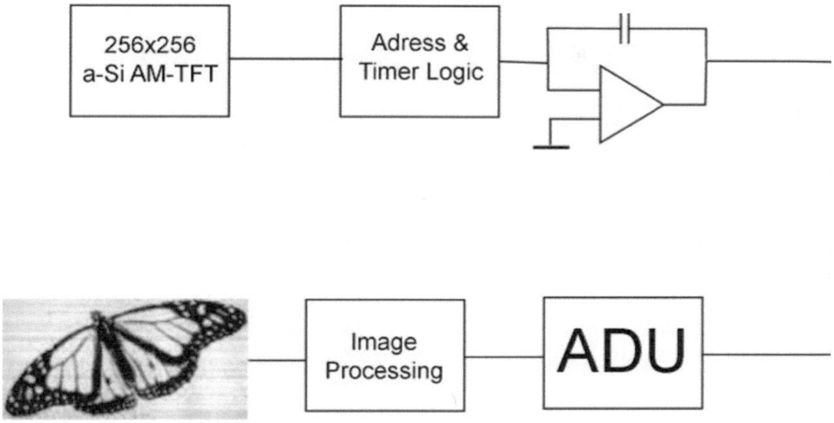

Figure 3.10.: Schematic for the read out electronics for imaging devices.

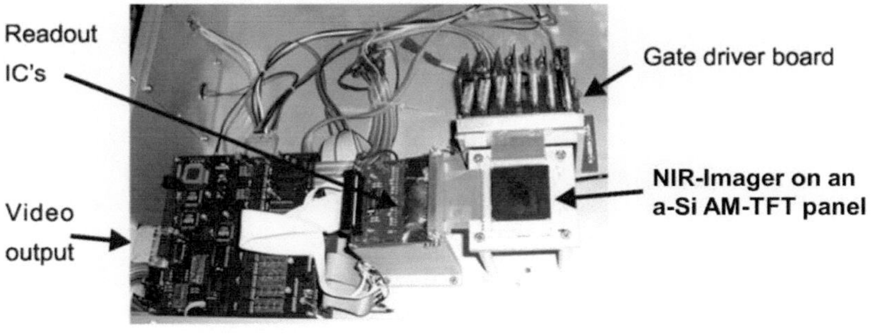

Figure 3.11.: Hardware realization of an ROIC for capturing images from an a-Si AM-TFT panel.

of shadow-casting through slides with individual designs is applied to create themes for an image on the imager's active area.

Methods for Image Reconstruction

Capturing images only, is not sufficient to receive proper images, so that corrections need to be made to the captured image. Post-processing images is done by flat-field correction (also known as flat-fielding). By flat-fielding, distortions in the optical path or inhomogeneous illumination by a light source as well as pixel fluctuations in sensitivity and dark currents can be corrected. For calibration, a dark and an illuminated image are needed in addition to the captured

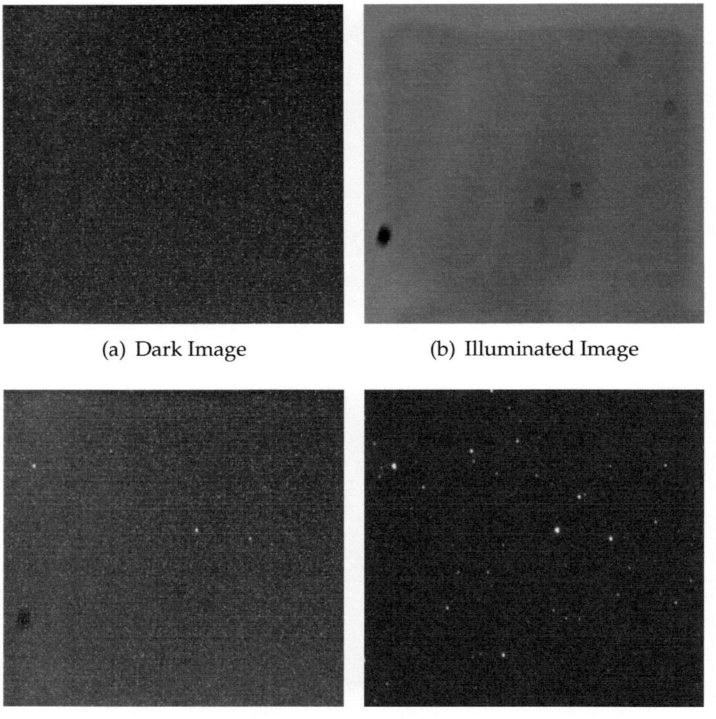

(a) Dark Image (b) Illuminated Image

(c) Captured Raw Image (d) Flat-Field corrected Image

Figure 3.12.: Flat-Field Correction. A dark (**a**) and an illuminated image (**b**) are captured ahead starting image acquisitions. According to equation 3.19 flat-fielding is applied to the raw image (**c**) and a corrected image (**d**) is obtained. The example was reproduced from `http://de.wikipedia.org/wiki/Bildsensorkalibrierung` at June, 2 2010

raw image (see figure 3.12). With the acquired image P_{acq} the following operations are performed.

$$P_{FF} = \frac{P_{acq} - P_{dark}}{P_{bright} - P_{dark}} \tag{3.19}$$

In the numerator offset structures are eliminated from the acquired image as well as from the illuminated image P_{bright}. Perturbation in the optical beam and fluctuations of the pixel sensitivity are considered by the division of the offset corrected image $P_{acq} - P_{dark}$ by $P_{bright} - P_{dark}$. In figure 3.12 the quality of the capture raw image of a starry sky can be dramatically improved by flat-fielding according to equation 3.19 (compare figure 3.12(c) and 3.12(d)).

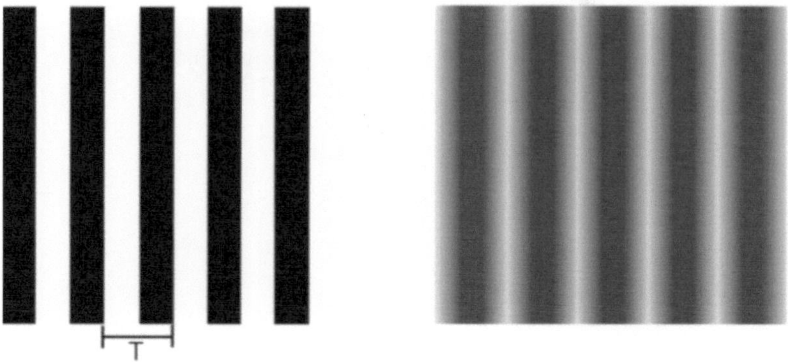

Figure 3.13.: Pattern for modulated transfer function measurements (**left**) and an expected reproduced pattern by the imager with fading the transition of the MTF pattern (**right**).

Resolution Capacities by Modulation Transfer Functions

The resolution of an imager is a further key performance indicator. To determine the imager's resolution, the modulation transfer function (MTF) is applied. According to figure 3.13 a periodic line pattern is created and projected on the imagers active area (here by shadow-casting). The periodic line pattern has a spatial frequency and the signal difference between the dark and the bright areas corresponding to the line pattern is observed. An ideal system with infinitely small pixels and without any scattering would resolve any line pattern independent of the spatial frequency (line pairs mm) as depicted in figure 3.14. However, in real systems when the spatial frequency reaches the domain of the pixel size, the pattern smears out between neighboring pixels and the transition from dark to bright is less promoted. Scattering of light is an additional reason for lower resolution (e.g. at encapsulation layers in the imaging device). In this work, the MTF is measured by applying lead coins with line patterns from x-ray imaging to the imager and the signal amplitudes below the patterns are analyzed. According to ref. [159] the MTF can be determined by

$$MTF(f) = \frac{A_{bright}(f) - A_{dark}(f)}{A_{bright}(f) + A_{dark}(f)} \tag{3.20}$$

with A_{dark} as the signal amplitude below the shadowed area and A_{bright} below the illuminated area of the line pattern with the spatial frequency f.

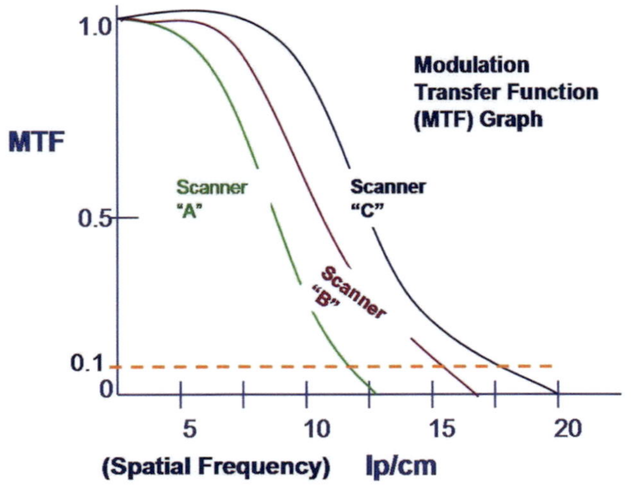

Figure 3.14.: The MTF of CT scanners as representatives for real imaging systems. In contrast to an ideal system the resolution performance decays with the spatial frequency of the line pattern. The graph was modified from `http://www.ceessentials.net/article33.html` on June, 2 2012.

3.1.4. Lifetime Analysis

For accelerated lifetime tests, standard testdiodes are kept at controlled climatic conditions. In a climate chamber (ESPEC SH-221) a temperature of 38°C with a relative humidity (RH) of 90% is applied. The lifetime measurements were developed for organic devices by Ptzold et. al [124, 123]. For such climate conditions an acceleration factor of $\sim$6 for lifetime test can be assumed [123]. As depicted in figure 3.15, I-V characteristics for standard testdiodes are acquired over time under accelerated lifetime conditions. The evolving I-V characteristics are validated in terms of photodiode performance like rectification and dark currents.

3.2. Structural Characterization Methods

Structural characterization is an essential part when designing new material composites and semiconductor devices. However, scanning electron microscopy (SEM) and transmission electron microscopy (TEM) methods have not been executed by our group, so that the introduction to SEM and TEM is kept to the working principles.

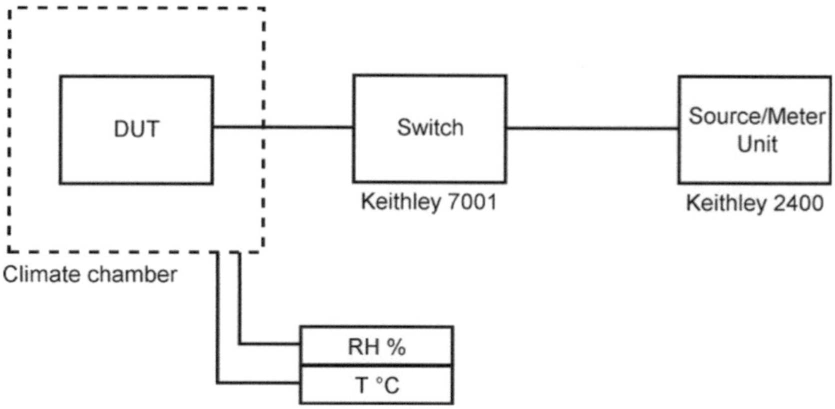

Figure 3.15.: Setup for accelerated lifetime measurements. The I-V characteristics of the DUT are measured over time at controlled climatic conditions. The conditions determine the acceleration factor. At 38°C and a RH of 90% the factor is ~6 [123].

3.2.1. Scanning Electron Microscopy

In an SEM an electron beam generated by emission from a thermionic cathode is bundled and focused by electromagnetic coils on the substrate (see figure 3.16). The electron beam scans the samples row by row and various signals can be detected due to the impinging electron beam on the samples. Besides the X-rays, back-scattered and secondary electrons are most commonly used to reconstruct information and images of the sample. Secondary electrons are emitted by the interaction of the electrons with the atoms and are characterized with low energies at a few eV. Thus, these electrons come from the first few nm of the sample, so that the topography of the sample can be determined by detection of the secondary electrons. Conclusions on the material distribution in the sample can be given by the back scattered electrons. The intensity is dependent on the atomic number of the materials and the electrons have energies of several keV. By detecting the back scattered electrons, a mapping of the materials in the sample (material contrast image) can be given, because the heavier elements generate more back scattered electrons than lighter materials.

3.2.2. Transmission Electron Microscopy

As shown in figure 3.16 TEM is a special operating mode of the electron microscopes. A sample is exposed to high accelerated electrons in a vacuum. By passing through the sample the electrons are scattered and absorbed, so that a

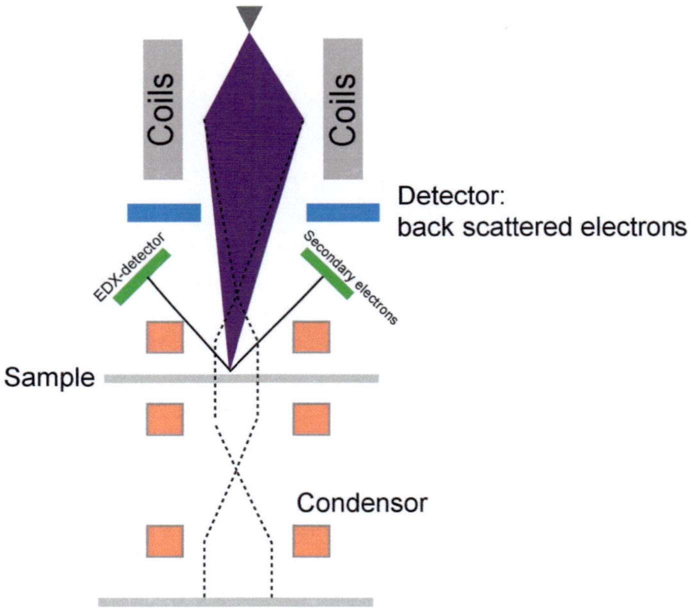

Figure 3.16.: Scanning and transmission electron microscopy.

bright field image can be captured by the absorption of the electrons in the film. Regions with either a thicker film or materials with higher atomic numbers appear darker. TEM is a powerful method with a variety of alternative characterization techniques as the visualization of crystalline structures by diffraction patterns. However, the TEM images in this work are bright field images. For detailed insight into electron microscopy see ref. [35].

3.3. Characterization Methods for Material and Blend Properties

3.3.1. Bulk Mobilities by Charge Extraction by Linear Increasing Voltage

With the CELIV method it is possible to conclude on bulk mobilities of charge carriers [71]. If a voltage ramp (see figure 3.17) in reverse direction is applied to extract charge carriers from the device, the current is composed by displacement and extraction currents: I_{dis} and I_{ext}. The displacement current is caused by the diodes capacity and can be described by

$$I_{dis} = C \cdot \dot{V} \tag{3.21}$$

Assuming full depletion of the diode due to the built-in potential, the capacity of the diode can be approximated by a simple parallel-plate capacitor which yields to

$$I_{dis} = \frac{\epsilon_r \epsilon_0 A}{d} \cdot \dot{V} \tag{3.22}$$

As a consequence of the constant capacitance, I_{dis} remains constant for linear increasing voltages. A differing current from I_{dis} can only be caused by the extraction current I_{ext} due to free carriers extracted by the electrical field of the applied voltage ramp. To avoid distortion of I_{ext} blocking contacts need to prevent injection of the corresponding charge carriers.

According to ref. [70] the conditions for low conductivity materials are given when the time constant of the diode is much smaller than the transit time of the charge carriers ($\tau_\sigma << t_{tr}$) so that $I_{dis} \geq I_{ext}$. Recently ref. [98] found that low conductivity conditions for CELIV measurements have to be considered more carefully. However, for the estimation of mobility within this work, the

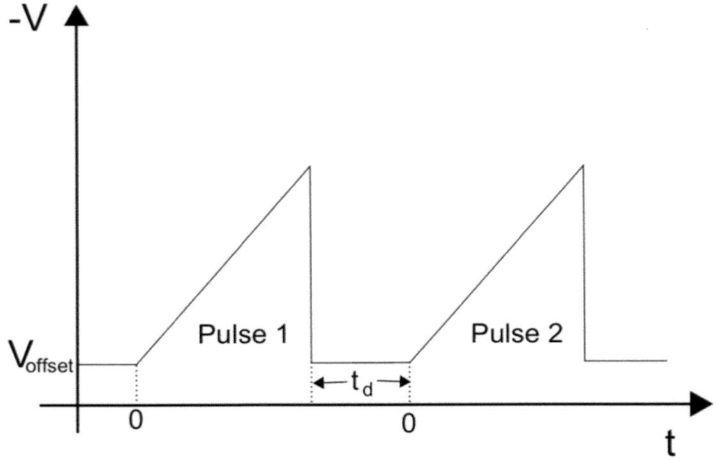

(a) Voltage ramp

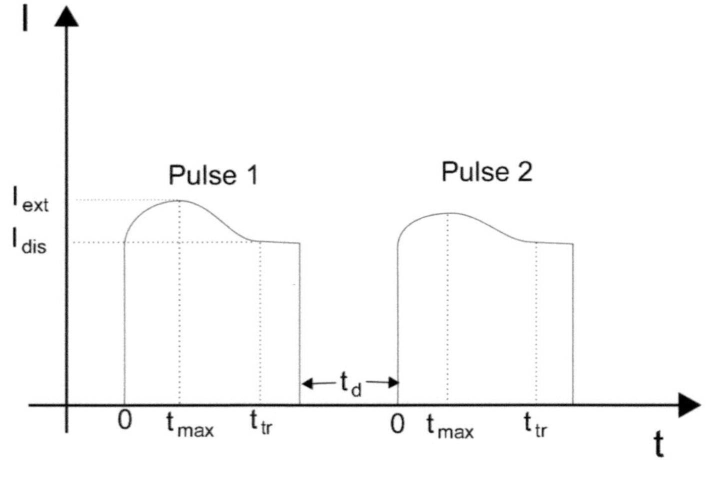

(b) Current response

Figure 3.17.: The charge extraction by linear increasing voltage (CELIV) method. (**a**) Two subsequent voltage ramps with a delay in between of t_d are applied to the diode. (**b**) A sketch of a common response of the diode to the applied voltage ramp. The current consists of two components: the displacement current I_{dis} and the extraction current I_{ext}. With the time (t_{max}), when I_{ext} reaches its maximum, conclusions can be made to charge carrier mobilities [71]. With the second pulse the recovery to equilibrium might be studied [71, 72]. At Photo-CELIV a light pulse is applied shortly before the voltage ramps.

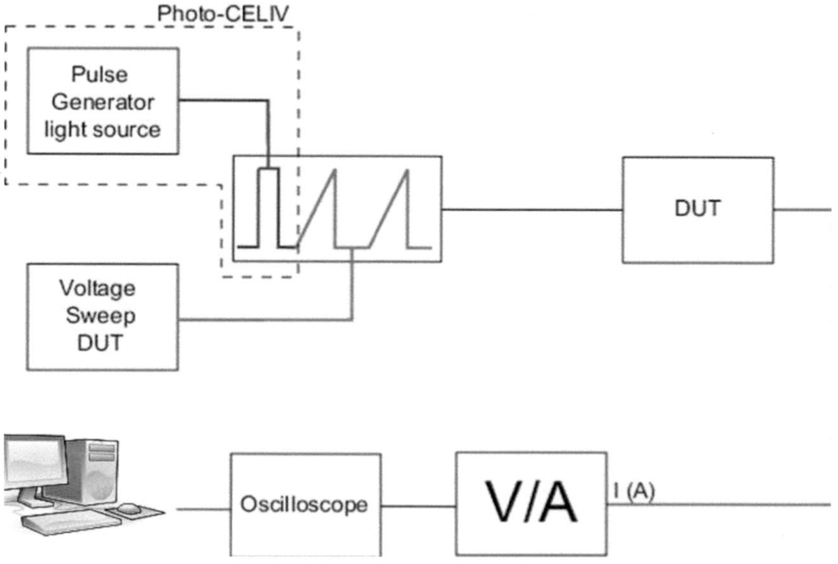

Figure 3.18.: Setup of a CELIV-measurement

method is still sufficient for determining and is given for the low conductivity conditions by ref. [72] by

$$\mu = \frac{2d^2}{3\dot{V}t_{max} \cdot \left(1 + 0.36 \cdot \frac{I_{ext}}{I_{dis}}\right)} \tag{3.23}$$

In figure 3.18 the setup is illustrated. The voltage sweeps are generated by a waveform generator (TTi TGA1242). The voltage offset and the current response are provided, amplified and converted by a current-voltage amplifier (Femto DHPCA100). The traces are captured and visualized by an oscilloscope (LeCroy Wavesurfer 424). By application of a light pulse briefly before the voltage ramps gives the Photo-CELIV.

3.3.2. Field Effect Mobilities

The analysis on field effect mobilities is straight forward. As known from semiconductor physics one can build field effect transistors from almost any semiconducting material (see ref. [171]). Thus, field effect measurements are a powerful tool to extract mobilities from films. A field effect transistor (FET) is made of a gate contact, which is electrically insulated from the semiconducting

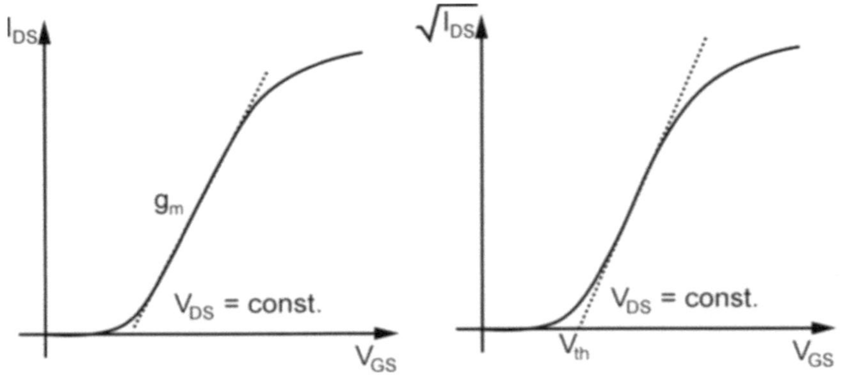

Figure 3.19.: Transfer characteristics and determination of parameters. (**Left**): In the linear regime (dashed line) of transfer characteristics the transconductance g_m can be extracted and correlates with the mobility. (**Right**): The threshold voltage V_{th} is determined by a line (dashed line) at the intersection with the x-axes.

bulk material by a dielectric (e.g. silicon oxide). Underneath or above the gate a source and a drain contact are buried in the bulk to inject the channel current I_{DS} which is controlled by the voltage between drain and source contacts V_{DS} and between gate and source contacts V_{GS}. The gate voltage repels or attracts charge carriers to interface between dielectric and semiconductor according to the strength of the electric field. The drain-source voltage drives the charge carriers through the channel. For a FET in saturation regime ($V_{DS} > V_{GS} - V_{th} > 0$) the current is given by

$$I_{DS} = \frac{\mu \cdot C_i}{2} \cdot \frac{W}{L} \cdot (V_{GS} - V_{th})^2 \tag{3.24}$$

The field effect mobility can be derived from the transconductance g_m which is described as

$$g_m = \frac{\partial I_{DS}}{\partial V_{GS}} = \mu C_i \cdot \frac{W}{L} \cdot (V_{GS} - V_{th}) \tag{3.25}$$

This leads to the relation for the mobility

$$\mu = \frac{\partial I_{DS}}{\partial V_{GS}} \cdot \frac{L}{W C_i} \cdot \frac{1}{(V_{GS} - V_{th})} \tag{3.26}$$

From this equation one can derive a reliable value of the mobility in the linear region of the transfer characteristics of a FET indicated on the left in 3.19. On the right in 3.19 the threshold voltage V_{th} can be extracted from $\sqrt{I_{DS}}$ to V_{GS} relation, which is the intersection of the V_{GS}-axis and a straight line through the intersection point and the corresponding slope. The channel length L and width

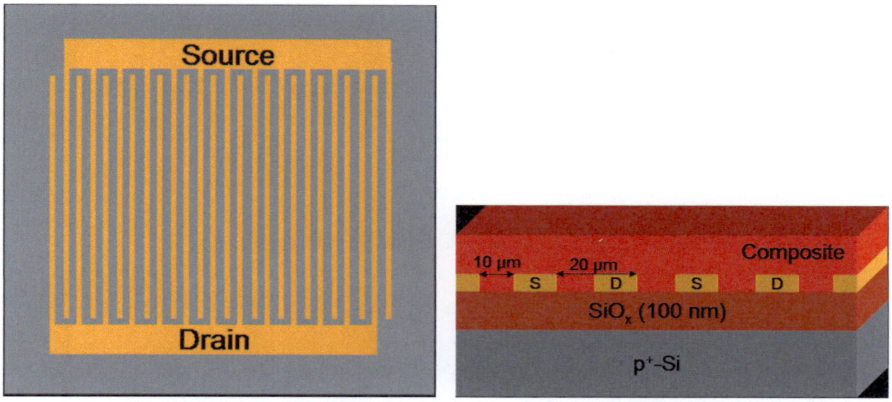

(a) Interdigitated Contacts

(b) Stack of FET

Figure 3.20.: Layout of backgated FETs. (**a**) An interdigitated structure for source and drain contacts was chosen to maximize the channel width of the FETs. (**b**) The cross-section through the FET illustrates the source (S) and the drain (D) contacts and their distance on top of a p^+-Si wafer with a thermally grown 100 nm SiO_x layer.

W are design parameters, whereas the intrinsic capacitance is approximated by an plate capacitance determined by $C_i = \epsilon_0 \epsilon_{ox} A/d$ where $A = LW$ is the area of the channel region. Standard FET-equations can be applied to organic materials in FET configuration as well (see ref. [60] and ref. [33]). In contrast to a common inorganic metal oxide semiconductor field effect transistor (MOSFET) forming an inversion layer, organic field effect transistors (OFETs) work in accumulation mode (majority charge carriers form the channel) as organic semiconductors show low conductivities in comparison to inorganic semiconductors. However, an OFET follows the equation introduced above and field effect mobilities can be extracted. OFETs within this work were fabricated on a 3″-p+-silicon wafer with a 100 nm silicon oxide on top. Contacts of a nickel (10 nm) as adhesion layer and 50 nm gold were sputtered and structured by a lift-off technique. To maximize the channel width W an interdigitated contact structure is used (see 3.20(a)). The channel length L and the size of the contact fingers is 10 µm (see 3.20(b)), so that the channel length W is 19.2 cm in total. The intrinsic capacitance C_i is 765 pF with a 100 nm of silicon oxide as gate oxide. Any voltage above 60 V has been avoided not to destroy the oxide as its dielectric strength is ~ 100 V.

After cleaning with acetone and isopropyl and oxygen plasma activation the semiconductor blend was deposited on the substrate by doctor-blading which was applied as well for the active layer of the photodiodes. According to the process of the photodiodes, the film was annealed for 10 min @ 80 °C. The

FETs were used in backgate configuration. Gate voltages were applied by a Keithley 2400 source-meter-unit and the channel currents I_{DS} were measured by a Keithley 236 source-meter-unit. All measurements on FETs were done in dark and under inert nitrogen atmosphere in a glovebox to minimize extrinsic influences.

3.3.3. Temperature Induced Conductivity

Temperature dependent measurements of I-V characteristics were made with the same setup in figure 3.1. With the Peltier element the temperature can be risen from 0°C to 80°C. Additionally, photoresponse measurements according to temperature dependence were conducted by the University of Linz at the Institute of Semiconductor and Solid State Physics. The device is placed in an optical cryostat for low temperature measurements. Additional, photoluminescence as well as spectral response measurements can be applied [127].

3.3.4. Transmission Spectroscopy

Actually, transmission spectroscopy is kept in this work at a pragmatic simple stage. The setup known from EQE measurements (c.f. figure 3.3) is modified in a purpose, that the sample (DUT) is placed in the beam right in front of the calibrated silicon or InGaAs photodiode which is located between the DUT and x/y/z translation stage. A reference measurement to calibrate the transmission measurement is conducted before. The monochromator sweeps the wavelength and accordingly a spectrum of the transmission is obtained by the relation of the reference signal to the measured transmission signal.

$$T = \frac{I_M}{I_R} \tag{3.27}$$

with T as the transmission, I_M as the measured current and I_R the reference current.

4. Materials, Sample Topology and Processing Technologies

Within this chapter materials used for the work, relevant technologies for processing and to manufacturing organic and hybrid devices are described.

4.1. Materials Used

4.1.1. Photoactive Layer Materials

Poly(3-hexyl-thiophene) (P3HT)

A polymer which is widely used and well investigated in organic photovoltaics is P3HT [58, 57]. It is characterized by high solubility, good processing and very suitable optoelectronic properties [17]. The structure of the regioregular (RR) P3HT is depicted in figure 4.1(a). Regioregularity of P3HT is described by head (H) to tail (T) bondings of the thiophene rings in the polymer. 100 % regioregularity is characterized by HTHT bondings throughout the polymer. RR-P3HT is superior to other configurations in terms of physical properties like charge transport, since HH-bondings deteriorate hole mobilities by an order of magnitude [122]. Furthermore, a high molecular weight of P3HT plays an important role for absorption and hole mobility. With higher molecular weight, a shift to the red in absorption and an increase in hole mobility can be observed [188, 48, 146]. RR-P3HT as an organic semiconductor has a HOMO of 4.9 eV and a LUMO of 3.0 eV and therefore a bandgap of 1.9 eV [155]. Since P3HT is a thoroughly investigated material, charge carrier mobilities have been already explored by several groups and methods: time-of-flight (TOF) [76, 122, 22], CELIV method [72], from current voltage characteristics [19, 48] and measurements on field effect devices [157, 24, 115, 177]. Depending on material properties originating from physical characteristics and processing, measurement method and conditions mobilities for holes from 10^{-5} $cm^2(Vs)^{-1}$ to 10^{-2} $cm^2(Vs)^{-1}$ can be found. In contrast to the common understanding that electron transport is negligible in P3HT (refer to ref. [115, 177]), ref. [24] and ref. [22] report a significant electron conductivity in P3HT with mobilities of 6×10^{-4} cm^2(Vs)$^{-1}$ and 1.5×10^{-4} cm^2(Vs)$^{-1}$. The glass temperature, which is important for temperature treatments of polymer films to influence film morphology, shows a broad variety in literature from 12 °C up to 110 °C. Note that the introduced RR-P3HT from Rieke Metals Inc. was exclusively used within this work. RR-P3HT from Rieke Metals is characterized by a regioregularity of more than 98 % and a molecular weight of 50 000 g/mol [65].

PCBM

PCBM is a derivate of the buckminster fullerene C_{60} which acts usually as an electron acceptor or conductor in organic devices, respectively. A fullerene molecule is able to carry up to 6 electrons [1]. However, PCBM is highly soluble in organic solvents achieved by the attachment of a methyl-ester group (see figure 4.1(b)). Electron mobilities for PCBM reported in literature range from 2×10^{-3} cm^2(Vs)$^{-1}$ to 2×10^{-2} cm^2(Vs)$^{-1}$ [110, 2, 115, 177]. Actually, ref. [2] found hole conducting properties of PCBM with high hole mobilities of 8×10^{-3} cm^2(Vs)$^{-1}$ in contrast to the prevailing conclusions of PCBM as electron conducting material with negligible hole mobilities [115, 177]. Energy bandgaps of PCBM reported in literature show values of -3.7 eV (HOMO) and -6.7 eV [155], while latest studies for PCBM gave -4.3 eV (HOMO) and -6.7 eV [160]. Note that the major role of PCBM is to separate excitons, withdraw electrons from the polymer and their extraction in a oBHJ device.

PCPDTBT

Poly[2,6-(4,4-bis-(2-ethylhexyl)-4H-cyclopenta[2,1-b;3,4-b′]dithiophene)-alt-4,7-(2, 1,3-benzothiadiazole)] (PCPDTBT) is a recently introduced low bandgap polymer for organic photovoltaic (OPV) with cyclopentadithiophene (see figure 4.1(c)) as a donor block [114] trying to push the absorption of polymers to the energetic optimum for efficiency in oBHJ solar cells to longer wavelengths. Charge carrier transport in this material is dominated by holes with mobilities in the range of 10^{-3} cm^2(Vs)$^{-1}$ to 10^{-2} cm^2(Vs)$^{-1}$ with a very weak contribution by electrons with a mobility of 10^{-5} cm^2(Vs)$^{-1}$ [112]. In comparison to P3HT the absorption is red shifted and has an absorption peak at 800 nm corresponding to an optical bandgap of 1.5 eV [114, 160]. Reported energy levels of PCPDTBT are -3.6 eV for the HOMO and -5.3 eV for the LUMO [114, 160, 112].

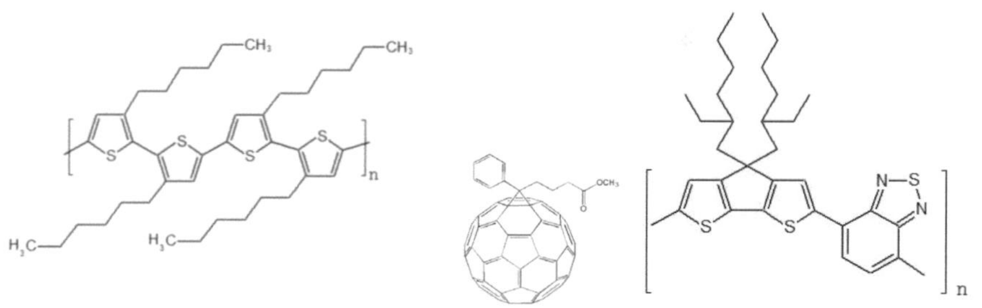

(a) Poly(3-hexyl-thiophene) (b) PCBM (c) PCPDTBT

Figure 4.1.: (**a**) Poly(3-hexylthiophene) (P3HT), (**b**) [6,6]-Phenyl-C61-butric-acid-methyl-ester (PCBM) and (**c**) Poly[2,6-(4,4-bis-(2-ethylhexyl)-4H-cyclopenta[2,1-b;3,4-b']dithiophene)-alt-4,7-(2, 1,3-benzothiadiazole)] (PCPDTBT)

Soluble, Colloidal Lead-Sulfide Quantum Dots (PbS-QDs)

In the theory chapter, quantum dots (QDs) have been introduced as three dimensional quantum confinements of particles (e.g. electrons) considered as artificial atoms. In contrast, quantum wells which are applied in LEDs and detectors the electrons can move freely in 2D space (2D electron gas) and are trapped only in the layer. QDs can be differentiated by core-only, type I, type II core-shell QDs and organically passivated QDs. Type I core-shell QDs are characterized by an inorganic semiconductor core with a smaller bandgap than the surrounding shell (e.g. ZnS/CdSe), so that both charge carriers experience a barrier from the core to the outer shell and are trapped on the core. This confinement is beneficial for the luminescence of those QDs. Usually, type I core shell QDs are capped by organic ligands for further solution based processing. Type II core-shell are slightly different. The inner core and the outer inorganic shell form an energetic heterojunction, so that either the hole or the electron is trapped on the core after excitation while the counter charge is transferred to the outer shell. A typical semiconductor QD, consists of an inorganic core, which is not larger than the Bohr exciton diameter of the corresponding bulk material, surrounded by an organic shell of ligands. The choice of a suitable capping ligand can be considered as a key point in an advanced colloidal synthesis of semiconductor QDs. Organic ligands regulate the growth rate and the size of QDs, can prevent them from oxidation and provide a dielectric barrier at the surface thus eliminating partially surface traps [56]. The present work especially concentrates on PbS-QDs (bandgap of bulk PbS: 0.41 eV) as for these QDs the tunability of PbS-QDs across the NIR regime (up to 2000 nm)is excellent and the synthesis is highly scalable for industrial appli-

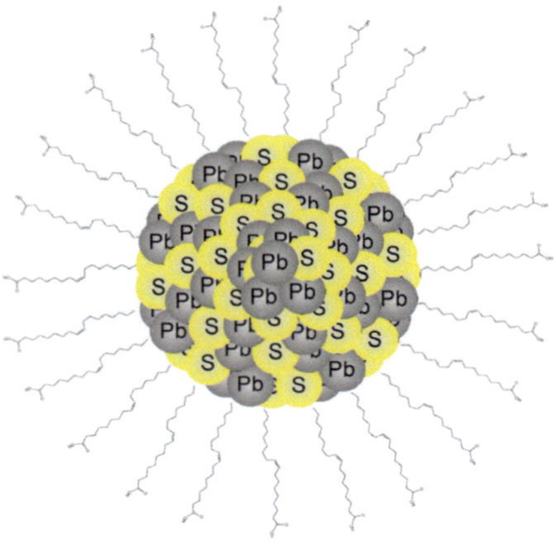

Figure 4.2.: Schematic drawing of an oleic acid capped lead sulfide quantum dot

cation due to inexpensive and safe precursor materials [55]. The conductivity of PbS-QDs are highly dependent on the applied ligand. PbS-QDs with butylamine ligands exhibit mobilities of 3×10^{-5} cm^2(Vs)$^{-1}$ and a conductivity of 4×10^{-7} Scm^{-1} [84] whereas with carboxylic acids (e.g. formic acid) mobilities of 2×10^{-3} cm^2(Vs)$^{-1}$ and a conductivity of 8×10^{-4} Scm^{-1} were found [186]. In contrast, oleic acid ligands insulate the PbS-QDs and do not show conductivity [86]. Note that the colloidal QDs used within this work are always organically passivated QDs and synthesized by the method from ref. [55].

4.1.2. Materials for HTLs

PEDOT:PSS

PEDOT:PSS is a conjugated polymer with a bandgap of < 2 eV. Figure 4.3(a) shows an oxidized PEDOT polymer, so that the PEDOT chain has a positive bipolaron. Many bipolarons create additional states in the bandgap and enhance the conductivity of the material. PSS anions neutralize the charges of the oxidized PEDOT molecules [51, 83]. PEDOT:PSS is usually deposited on top of an ITO layer to enhance injection (for light emitting diodes) or extraction properties (for solar cells). Since thin films below 150 nm only are coated on ITO, the electrode remains highly transparent in the visible. The PEDOT:PSS used in this work has an work-function of about 5.2 eV, a PEDOT:PSS ratio of

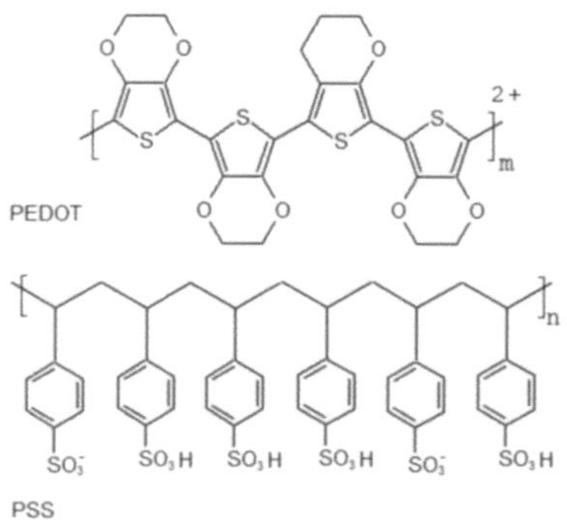

(a) PEDOT:PSS

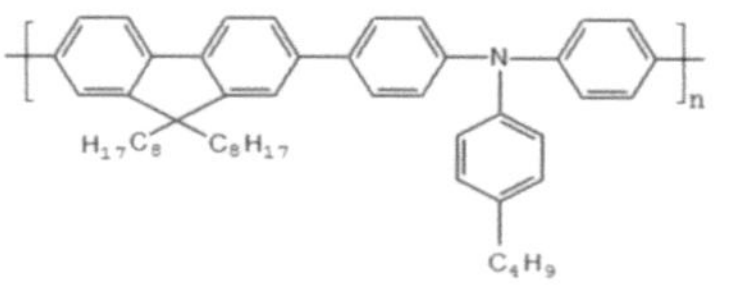

(b) TFB

Figure 4.3.: Chemical structure of (**a**) PEDOT:PSS and (**b**) TFB

1:20, a resistivity of 1×10^5 to 3×10^5 Ωcm and was purchased from H.C.Starck (CLEVIOS P VP CH 8000) now Heraeus Clevios GmbH [54].

TFB

Poly[(9, 9-dioctylfluorenyl-2, 7-diyl)-co-(4, 4'-(N-(4-sec-butylphenyl))-diphenyl -amine)] (TFB) is a hole conducting conjugated polymer (see figure 4.3(b)), which has energetic levels of 5.3 eV (HOMO) and 2.1 eV [21]. In OPDs this polymer is typically used as interlayer to reduce leakage currents by suppressing injection of electrons from the anode at reverse bias conditions. In OLEDs TFB is applied as electron blocking layer confining the electrons at the emitting layer for an enhanced optical recombination.

4.1.3. Materials for Electrodes

Indium Tin Oxide (ITO)

ITO is widely used as transparent conductive oxide (TCO) and especially in OPV devices as anode material. ITO is composed of 90 % indium oxide (In_2O_3) and 10 % tin(IV) Oxide (SnO_2). SnO_2 creates the necessary defects in the lattice of In_2O_3 leading to resistivities between 10^{-4} Ωcm to 10^{-3} Ωcm. -4.7 eV is the work function of ITO found by [28, 20]. Deposition is commonly done by

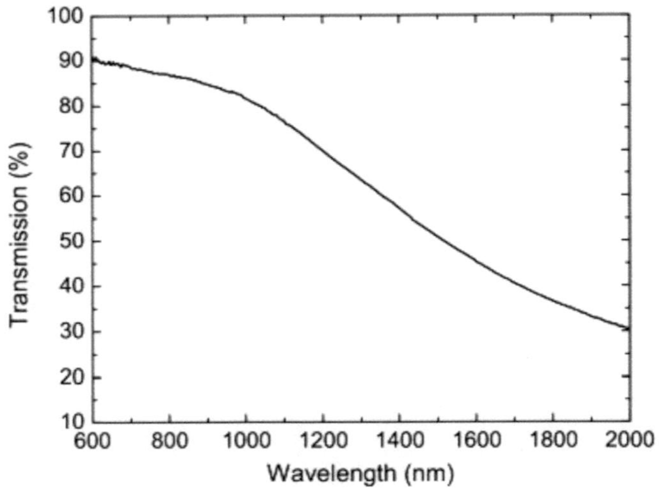

Figure 4.4.: Transmission of the commercial ITO substrate in the Vis-NIR spectra.

sputtering, but thermal evaporation, spray pyrolysis, pulsed magnetron sputtering and printing are alternatives to coat substrates with ITO films [46]. Commercially available ITO coated substrates with a 130 nm film with resistivity of 13 $\Omega/^2$ were used for device fabrication. ITO substrates have proven high transparency in the visible of more than 90 % and show adequate transmission for NIR radiation from ~80 % at 1000 nm and ~70 % at 1200 nm to ~30 % at 2000 nm (c.f. figure 4.4).

Cathode and Further Electrode Materials

An organic photodiode as discussed in section 2.3.3 needs cathode materials with low work functions to create a built-in potential to extract photogenerated charge carriers at low leakage currents.

Calcium (Ca) Calcium (Ca) is characterized by a very low work function of 2.87 eV [108] and a melting point of 843°C [50]. Therefore Ca is well suited for the design of organic photodiodes. However, as a consequence Ca is very sensitive to ambient conditions as Ca is oxidized immediately.

Silver (Ag) The noble metal Ag is often used as protective layer for low work function metals (e.g. Ca) to prevent oxidation and to enhance conductivity. To investigate single charge carriers (holes) in organic devices the counter electrode to ITO can be made of Ag which has an work function of 4.7 eV and is deposited by thermal evaporation (melting point: 962°C) [50].

Aluminum (Al) Aluminum (Al) is an established metal for electrodes in the whole electronic industry. With its work function of 4.28 eV [108] in comparison to 4.7 eV of ITO, Al is also used as a cathode in organic photodiodes and deposited by thermal evaporation (melting point: 660°C [50]).

Lithium Fluoride (LiF) Lithium Fluoride (LiF) is a salt in crystalline form and is used to modify the work function of an Al cathode. With a 3 nm layer of LiF the work function of an Al electrode can be reduced from 4.2 eV to 2.5 eV [149]. LiF is deposited by thermal evaporation and has a melting point of 660.4°C.

Gold (Au) Contacts from Au are made in this work to investigate single carrier devices and to fabricate OFETs. Au is characterized by a workfunction of 5.1 eV [108] and a melting point of 1064°C [50].

4.2. Manufacturing Technologies

The processing technologies can be restricted on layer deposition, since the photodetector devices are built-up layer by layer. The layers are made from wet and dry deposition and structuring processes.

4.2.1. Wet Chemical Deposition and Structuring

To deposit material from solution the following four processes have been used which are well known from inorganic semiconductor technologies. In addition, wet chemical structuring by lithography is shown, too.

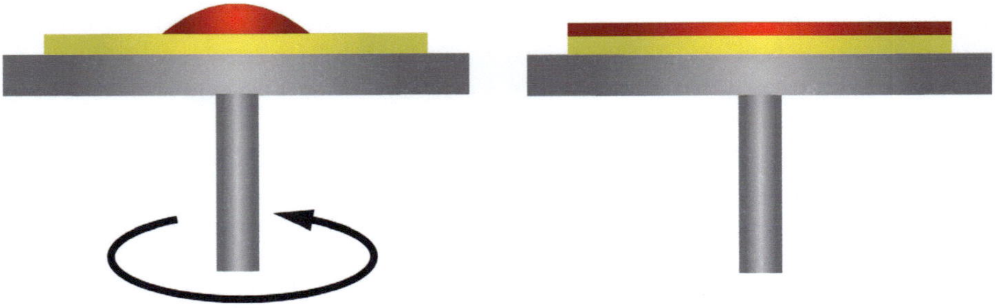

Figure 4.5.: Spin-coating: Deposition from solution on a substrate by rotation

Spin-Coating

Spin-coating is a very common method to generate homogeneous layers from solutions (e.g. photoresists). A substrate is placed on a chuck and fixed to it by a vacuum. A defined volume of the solution is dispensed on the substrate and spun at a certain velocity (see figure 4.5). The acceleration, spin velocity and spin time are the most important parameters of the apparatus for film formation, while surface properties and viscosity are determined by the substrate and the solution, respectively. A more detailed description in terms of practical application and theoretical background are given by ref. [27, 152].

Doctor-blading

The doctor-blading technique is an alternative to generate films from solutions. In contrast to spin-coating a doctor blade moves with a specified distance over the substrate (see figure 4.6). With the distance and the velocity film thicknesses can be controlled keeping in mind that viscosity of solutions and surface properties of the substrates have strong influences as well on the film. In general, doctor blading can deal with larger areas and thicker films as by spin coating process.

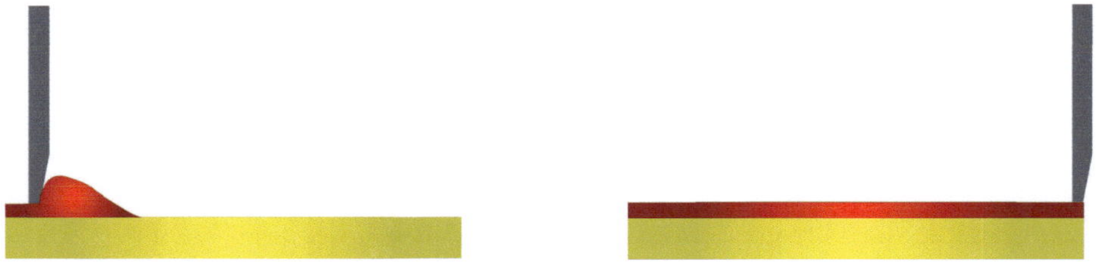

Figure 4.6.: Doctor-blading technique for film deposition from solution

Spray-Coating

Another option for large area processing from solution is the spray-coating technique. Spray-coating is already widely used in painting like airbrushing cars, etc. and can be easily applied to solution processable semiconductor materials. The working principle is the same. With an airbrush the solution is sprayed on the substrate. Here, the substrate moves on an x/y-stage underneath the spraying cone (see figure 4.7). Important parameters are the distance of the airbrush to the substrate, the velocity of the movements of the x/y-stage, the size of the spraying cone and the concentration / viscosity of the solution. Spray coating is superior when depositing thick layers and has the big advantage to be applied multiple times on the same substrates. Spray-coating reduces dissolving the underlying layers compared to spin-coating and doctor-blading.

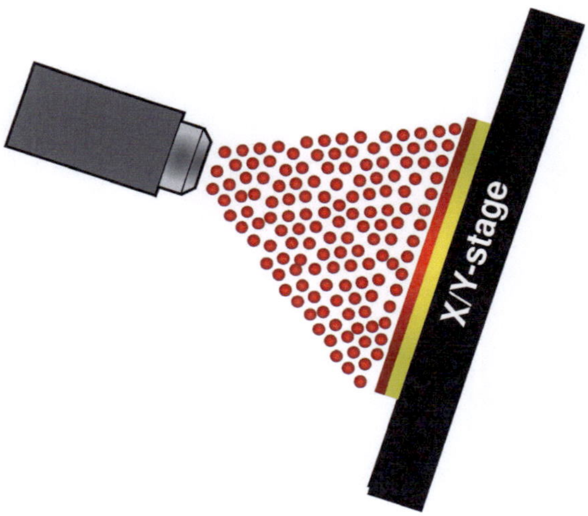

Figure 4.7.: Spray-Coating: A technique known from painting (airbrush) applied to solution processable organic semiconductors

Drop-casting

For completeness reasons the drop-casting method is also mentioned, since this method was used sometimes in this work. From e.g. a syringe a drop with a defined volume is dispensed on the substrates. This method is mainly used, when one has to handle small amounts of rare materials. However, the controllability and reproducibility are fairly poor for the films and materials in this work.

Lithographic Structuring

Some layers of the devices need to be structured. Lithographic structuring is an opportunity to create patterns. The procedure is shown in figure 4.8 for positive and negative photoresists. A photoresist is spun on the layer to be structured. The photoresist is illuminated through a mask with UV light (here: 365 nm). After development the illuminated part is removed in case of a positive photoresist whereas the illuminated part remains when a negative photoresist is used. Because of the protection of the photoresist the opened domains are wet chemical etched and the layer obtains the designated pattern.

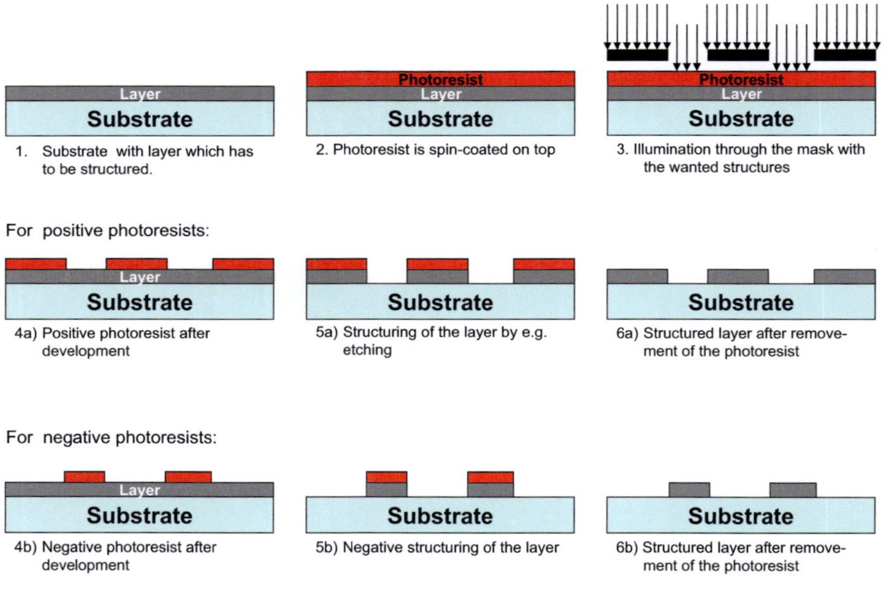

Figure 4.8.: Lithographic structuring of layers with positive and negative photoresists

Lift-off

To avoid etching a lift-off can be performed to structure the layer. In general, the structuring is very close to the procedure mentioned before. The whole process is shown in figure 4.9 and the differences are discussed. On the substrate and underneath the photoresist a layer is deposited to improve the undercut formation. After development the pattern defines where the layer is deposited on the substrate whereas the material on the photoresist is removed during lift-off. The undercut facilitates the removal of the designated parts.

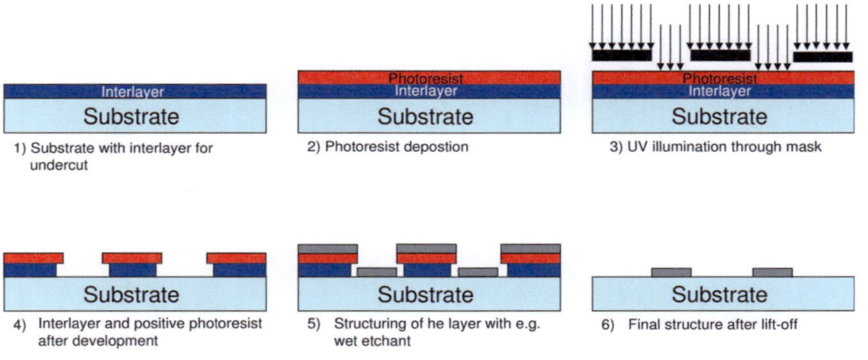

Figure 4.9.: Lift-off process with a positive photoresist

4.2.2. Physical Vapor Deposition Technologies

To deposit electrodes two methods, thermal evaporation and sputtering were applied, which belong to the family of physical vapor deposition.

Thermal Evaporation

Thermal evaporation is in principle a very simple process. In a chamber with ultra-high vacuum (usually 10^{-6} mbar to 10^{-7} mbar) a boat or a filament is heated up until the material (e.g. Al, Ag, Ca) evaporates and condenses on the substrate (see figure 4.10. With the heat source (here: current through the boat) the evaporation rate can be controlled while thickness and deposition rate is monitored by a oscillation quartz sensor. This method is a good choice, when gentle impact on the surface is wanted which correlates with the evaporation rate. The ultra-high vacuum is needed to avoid interactions between the evaporated material and impurities, since in OPDs highly reactive materials as Ca,

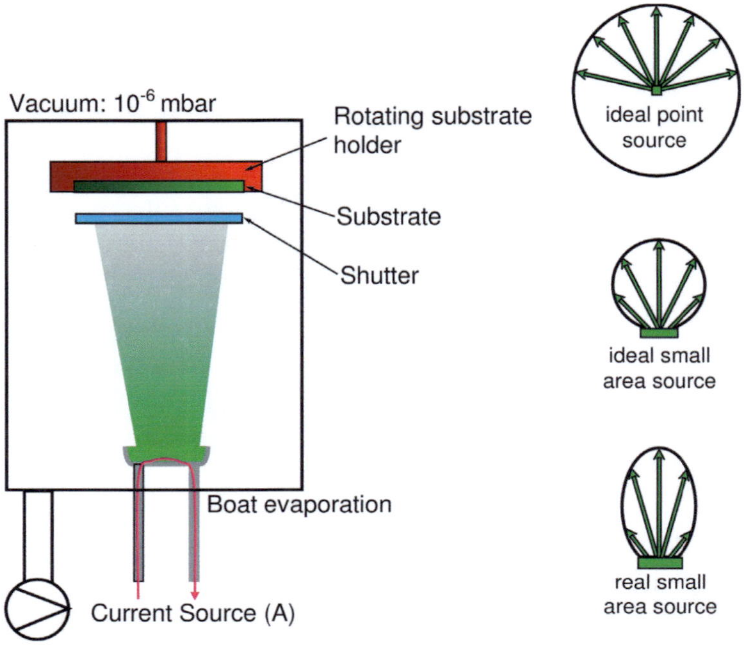

Figure 4.10.: Thermal evaporation in an ultra-high vacuum chamber.

LiF, Mg, etc. are used as top contacts. Furthermore evaporation has very low divergence of the evaporation direction. Therefore, edges will not or will be only slightly covered.

Sputtering

Sputter deposition is enabled by creating plasma between to electrodes in a ultra-high vacuum chamber. The plasma is usually created by an inert gas flux (e.g. Argon). When a bias is applied to electrodes the intrinsic electrons and Argon ions are accelerated and further neutral Argon atoms are ionized by impacts of electrons and ions. The ignition of the plasma is easily visible as it shines blue / violet in case of an Argon plasma (control by eye) since impact processes transfer energy to the atoms (excited states / electrons) which relax radiatively. Electrons and Argon ions are driven further to the corresponding electrodes. The material which needs to be deposited is fixed to the negative electrode (target), because the impacts of Argon ions ablate material which condenses on the substrate at the positive electrode. Inversing this configuration plasma etching can be done. So far, DC bias was applied only and thus solely metal can be sputtered, whereas for insulators the target is charged which interrupts the sputter process. To sputter insulators a high frequency generator is used instead of a DC bias source. In an sufficiently fast alternating electric field Argon ions cannot follow the field anymore, while the electrons do. The sputter equipment is designed, that the capacitor between plasma and target is much smaller than the capacitor between substrate and plasma. Therefore, when small amounts of charges are shifted, a much higher voltage is created between target and plasma. Thus, the Argon ions are accelerated to target. In that manner, insulators can be deposited by sputtering. Due to the scattering with the plasma, the deposition of the material is very homogeneous (better covering of edges), while the impact is of higher energy compared to evaporation and the plasma can affect sensitive films like organic films.

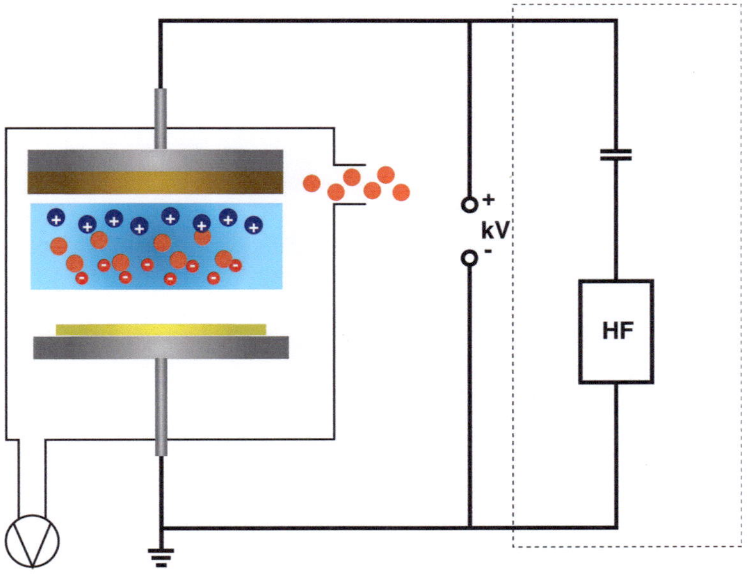

Figure 4.11.: Process of sputtering in ultra-high vacuum.

4.3. Preparation Methods by Device Layout

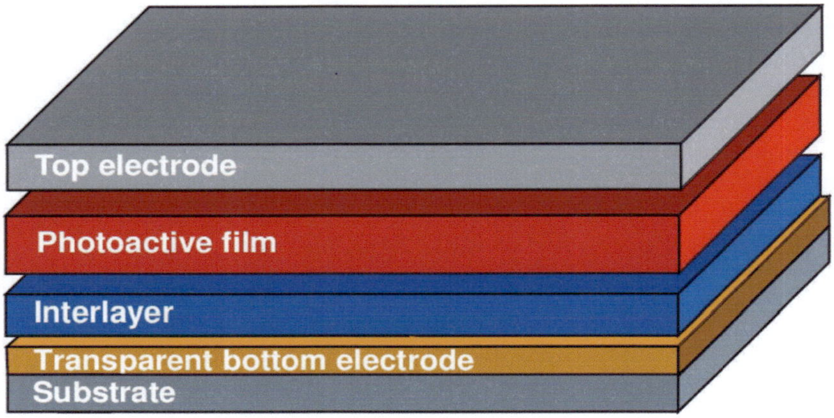

Figure 4.12.: Schematic layout of an organic and hybrid photodiode

In this section the standard and different modified preparations of organic and hybrid photodiodes are discussed following the layout shown in figure 4.12. The standard methods are stated in addition with alternative used techniques. Figure 4.12 gives the stack of a standard device. The device is built up on a substrate (usually glass) with a transparent electrode on top, followed by an interlayer. Afterwards, the photoactive layer consisting of organic and inorganic semiconducting materials is deposited. To end with, a metal contact is put on top. Not shown in figure 4.12 is a hermetical sealing of the devices to avoid influences and degradation from ambient conditions.

4.3.1. Bottom Electrode

As already discussed in section 2.3.3 an OPD needs a transparent and a high work-function electrode for adequate device operation and properties. Up to date, ITO is widely and commonly used for this purpose. ITO is fairly transparent in the visible and exhibits as well an adequate transmission in the NIR, so that this material was selected as bottom electrode. The ITO coated glass substrates (120 nm ITO) used within this work were purchased from the Merck company. Bottom electrodes were structured as described in figure 4.2.1 by lithography and wet etching. As wet etchant a warm bath (48°C) of hydrogen bromide (HBr) was applied for several minutes.

4.3.2. Hole Transporting Layer

Subsequently to the bottom electrode an interlayer is deposited on the substrate to improve hole extraction and the semipermeable behavior of the contacts. PEDOT:PSS is the standard choice and is spin-coated achieving roughly 100 nm. Note that for photodiodes the low conductive alternative CH8000 from H.C. Starck is used to improve dark current properties. Different interlayers are monitored and are indicated in the text.

4.3.3. Photoactive Layer

The photoactive layer is processed from solution and consists in case of an OPD of a conjugated polymer (here: P3HT or PCPDTBT) and a fullerene derivate (here: PCBM). The components are brought into solution with chlorobenzene (roughly 2 wt%) or xylene (1.75 wt%). The ratios in weight are stated appropriately when discussing the results.
For IO-HPDs all components of the photoactive layer (generally PbS-QDs, P3HT and PCBM) are dissolved with 2 wt% in chlorobenzene separately and are blended together right before deposition in a particular ratio given as well individually in the respective "'result"' part of the thesis.
For all the semiconductor solutions treated within this work the deposition methods drop-casting, doctor-blading and spray-coating are applied and compared later in terms of device performance.

4.3.4. Electron Transporting Layer

As already depicted in figure 4.12 in the devices no electron transport layer (ETL) or hole blocking layer (HBL) is usually considered. However, the application of PCBM for such a layer is discussed later in chapter 6.2 without having an impact on increasing device performance.

4.3.5. Top Electrode

Top electrodes are chosen to be of lower work function to generate a built-in potential with the high work function bottom electrode (see figure 4.1.3). Extensively used in this thesis are Ca/Ag (Ag is for protection of the highly reactive Ca) and Al top electrodes which are deposited by thermal evaporation. Note that during optimization and experimental work more alternative top electrodes were evaluated.

4.3.6. Encapsulation

Since the organic and hybrid composites as well as the electrodes are sensitive to ambient conditions, the devices are hermetically sealed by an encapsulation with an UV and heat curing adhesive and an encapsulation glass.

5. Organic Low Bandgap Polymer Photodiodes for Near-Infrared Detectors

Near-infrared sensitivity of organic optoelectronic devices evoke many research activities since organic photovoltaics needs to extend the harvesting of photons of solar light from the visible to longer wavelengths. A very common stack of organic photovoltaics and photodetection is a blend of P3HT:PCBM. In figure 5.1 the characteristics of a state of the art device in the visible and the accomplished I-V characteristics and quantum efficiencies are shown for doctor-bladed samples without any further treatment. Suitable organic photodiodes for the near-infrared need to show similar characteristics in terms of dark current densities and efficiencies. As P3HT:PCBM films show low dark currents of $\sim 3 \times 10^{-3}$ mAcm^{-2} (lower dark currents by spray-coating techniques demonstrated: $\sim 5 \times 10^{-5}$ mAcm^{-2} [176]) and EQEs of almost 80% (figure 5.1(b)). The challenge of this work is to shift the spectral sensitivity to the NIR by preserving the good performance of P3HT:PCBM photodetectors. Note that the requirements of photodetector application are not exclusively linked to dark currents and spectral responsivity, since noise characteristics have also a strong impact. Signal resolutions of detectors are defined by the NEP (see paragraph 3.1.2) whereas for photovoltaics the power conversion efficiencies are important. A basic approach is to exchange the visible absorber of P3HT with a material absorbing at longer wavelength. Recent progress mostly from organic photovoltaics made a variety of potential polymers available for near-infrared detectors. Dithiolenes (figure 5.2(a)) were one of the first organic materials demonstrating NIR response [117]. In particular, modifications of polythiophenes show promising properties for NIR sensitivity illustrated among other polymers with reported NIR sensitivity in figure 5.2(b) to 5.2(f). An overview of achieved properties of the molecules is given by table 5.1. Encouraging results for longer wavelengths (800 nm to 1000 nm) are obtained by PTBEHT and by LBPP-1 as they show efficiencies of 18% and 10%, respectively. PCPDTBT and PTT exhibit even competitive EQEs of $\sim$35% and $\sim$44%, respectively to P3HT devices.

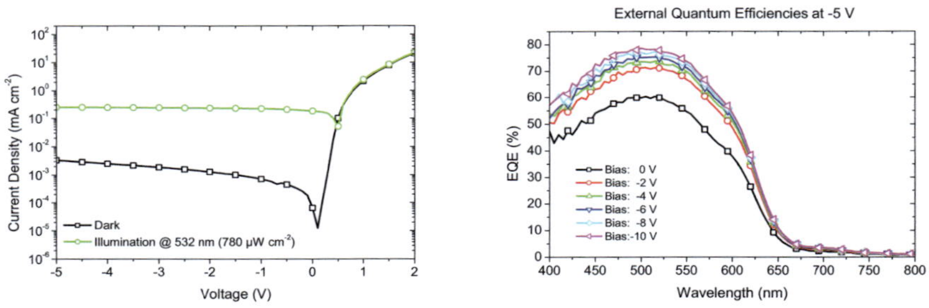

(a) Current Densities

(b) External Quantum Efficiencies

Figure 5.1.: State of the art photodetectors of doctor-bladed P3HT:PCBM for the visible range up to 650nm. (**a**) I-V characteristics in dark (black squares) and under green illumination at 780 µW cm^{-2}(green circles). (**b**) EQE for different reverse biases.

Polymer	Abbre-viation	NIR absorption range	Peak Efficiency	Dark currents (mAcm^{-2})	Ref.
poly[2,6-(4,4-bis-(2-ethylhexyl)-4H-cyclopenta[2,1-b;3,4-b′]dithiophene)-alt-4,7-(2,1,3-benzothiadiazole)]	PCPDTBT	700-800 nm	35% @ 0 V	10^{-3} @ -5 V	[63, 160, 114]
Polythieno[3,4-b]thiophene	PTT	650-900 nm	44% @ -5 V	10^{-2} @ -2 V	[184]
poly{5,7-bis[3,4-di(2-ethylhexyloxy)2-thienyl]-2,3-diphenyl-thieno[3,4-b]pyrazine}	PBEHTT	$<$ 900 nm	few % @ 0 V	10^{-1} @ -1 V	[182]
poly{5,7-di-2-thienyl-2,3-bis(3,5-di(2-ethylhexyloxy)phenyl)-thieno[3,4]-bypyrazine}	PTBEHT	800-1000 nm	18% @ 0 V	10^{-1} @ -1 V	[182]
	LBPP-1	900-1100 nm	10% @ 0 V	10^{-1} @ -2 V	[126]

Table 5.1.: Potential near-infrared sensitive polythiophene derivatives with reported characteristics.

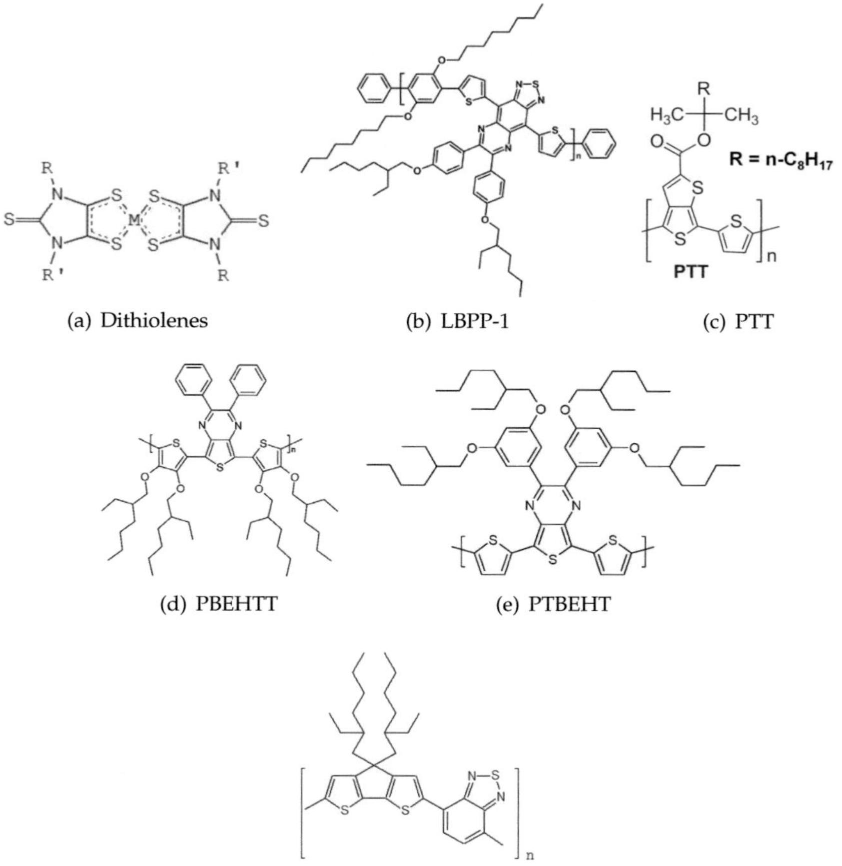

(a) Dithiolenes (b) LBPP-1 (c) PTT

(d) PBEHTT (e) PTBEHT

(f) PCPDTBT

Figure 5.2.: Selected polymers with NIR sensitivity. a) Dithiolenes as a first pioneer and emerging polythiophene derivatives with promising characteristics for photodetection b-f).

In the following further investigations on efficient near-infrared sensitive photodiodes based on PCPDTBT are performed, since they proved to operate at low dark currents[1], whereas the other polymers showed at least one order of magnitude higher dark currents.

5.1. Donor/Acceptor Ratio in PCDTBT:PCBM Films

Organic photodiodes are characterized by bulk-heterojunctions with polymer / fullerene composites as donor/acceptor systems for efficient photogenerated charge carrier extraction [143, 185]. A composition of a PCPDTBT:PCBM blend was chosen for further optimization, since this blend is closely process compatible to P3HT:PCBM devices and as stated in table 5.1 has very promising properties. A concentration sweep of the electron acceptor with reference to the low band gap polymer is conducted to find and to approve the optimum composition for the blend for the current device processing conditions.

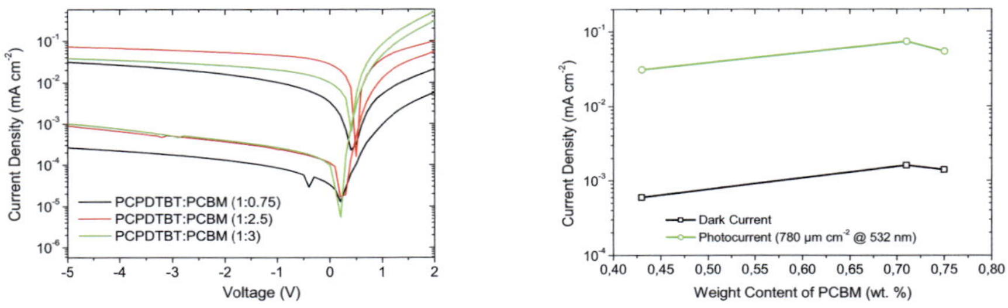

(a) Current Densities

(b) Current densities at −5 V

Figure 5.3.: (**a**) I-V characteristics of different ratios of PCPDTBT:PCBM in dark and with green illumination (780 µW cm^{-2} @ 532 nm) and (**b**) the dependence of dark and light currents on the PCBM weight content

In figure 5.3(a) and in figure 5.4(a) the I-V characteristics and the spectral sensitivity of doctor-bladed PCPDTBT:PCBM devices with reference to a P3HT:PCBM blend of a weight ratio of 1:1 and an active layer thickness of 200 nm are depicted. PCPDTBT:PCBM films with low PCBM contents exhibit low dark currents (3×10^{-4} mAcm^{-2} at -5 V) and low light currents (3×10^{-1} mAcm^{-2} at -5 V) under green illumination at 532 nm and

[1]Personal communication with Christoph Brabec / Konarka Technologies

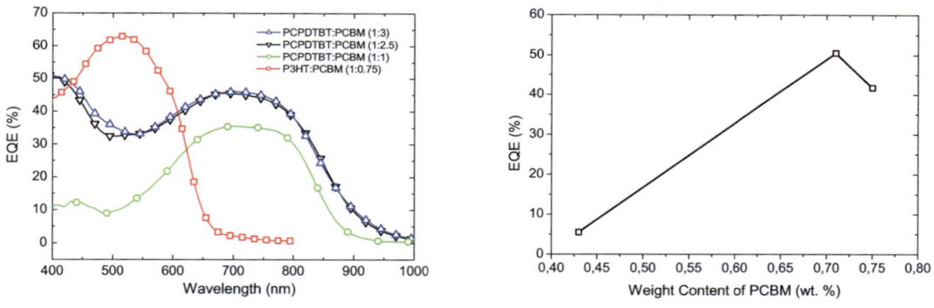

(a) EQE at -5 V

(b) EQE at 725 nm and -5 V

Figure 5.4.: Spectral sensitivity of PCPDTBT:PCBM with different PCBM contents. (**a**) Spectral sensitivity of devices with different PCBM contents and a P3HT:PCBM reference samples. (**b**) EQE at 725 nm with increasing weight content of PCBM

780 µW cm-2. Increasing the PCBM content gradually increases dark currents ($\sim 10^{-3}$ mAcm-2 at -5 V), though significant higher photoresponses (5.4×10^{-1} mAcm-2 to 7.3×10^{-1} mAcm-2 at -5 V) are observed. In forward direction, a significant rise in the current is observed when increasing the PCBM content (figure 5.3(a)). The serial resistance determines the behavior in forward direction. Consequently, with growing PCBM contents the resistivity of the film is reduced significantly. During illumination resistivity is further decreased due to photodoping of the film. The dependence of dark and light currents at -5 V on the PCBM weight content is depicted in figure 5.3(b). To conclude, a sufficient high PCBM content is necessary for an efficient charge transport in PCPDTBT:PCBM blends. The use of the electron acceptor PCBM in excess is mandatory to provide a balanced electron/hole transport. This necessity of high PCBM content blends is observed as well from EQE measurements as these devices show best efficiencies in the NIR of $\sim$46% at -5 V in contrast to $\sim$35% at -5 V for a 1:1 ratio (see figure 5.4(a)). In figure 5.4(b) the EQE is plotted versus the weight content of the PCBM in the film and a trend to higher PCBM content is clearly observable. Thus, it is obvious that separation and extraction of the photogenerated charge carriers benefit from high contents of PCBM. The 1:3 ratio has external quantum efficiencies of $\sim$46%, whereas the 1:2.5 ratio reaches $\sim$45%.

According to ref. [63], at an optimum ratio of 1:3.3 the lifetime of photogenerated carriers increases, which is also confirmed by solar power conversion efficiency measurements with 1:3 ratios by ref. [114]. Thus, the use and excess of electron acceptors seems to be necessary to provide a balanced and efficient charge transport in PCPDTBT:PCBM films [112]. The better injection and transport properties in forward direction for high PCBM content devices can be assigned to this effect. High PCBM content blends of 1:2.5 and 1:3 show comparable characteristics. However, the best EQE of 46 % is measured for the 1:3 blend in this work and in accordance to reported results of other groups, this ratio of 1:3 is kept constant for further investigations.

5.2. Deposition Techniques and Post-Annealing

The effect of deposition techniques of doctor-blading and spray-coating on the performance of P3HT:PCBM devices have been demonstrated by ref. [176]. Therefore, spray-coated and doctor-bladed PCPDTBT:PCBM samples will be compared within the following paragraph. In figure 5.5(a) a spray-coated sample is put in contrast to a doctor-bladed sample in figure 5.5(b). Both deposition techniques are suitable for PCPDTBT:PCBM (1:3) films as they only differ slightly in dark currents (black squares in figure 5.5) from 5×10^{-4} mAcm^{-2} to 2×10^{-4} mAcm^{-2} for the spray-coated and the doctor-bladed device, respectively. No difference at all is given for green light currents (780 µW cm^{-2} @ 532 nm) with values of 10^{-1} mAcm^{-2} at -5 V. Modification of performance by post-annealing is known from literature for P3HT:PCBM devices [59, 121], especially dark currents can be decreased [176]. For both, doctor-bladed and spray-coated PCPDTBT:PCBM devices a reduction of the dark current by post-annealing can be observed in figure 5.5. With two subsequent steps of additional thermal treatment after processing at 100°C and 140°C for 30 min a further reduction of dark currents to $\sim 7 \times 10^{-5}$ mAcm^{-2} and to $\sim 5 \times 10^{-5}$ mAcm^{-2}, respectively was possible for the spray-coated and the doctor-bladed devices while keeping the photoresponse constant. For low reverse voltages, the extraction properties were deteriorated. That means, that either hole or electron transport properties of the film or charge carrier extraction at the interfaces are affected by thermal annealing and results in a lower $\mu\tau$ product of the charge carriers (see section 2.3.3). A lowering of the mobilities is assumed as the conductivity is $\sigma \propto \mu$. This can be seen in the forward direction as with post annealing the forward current density is reduced and therefore the resistivity of the device increases. Both deposition effects suffer from this effect, so that post-annealing influences the film properties. The very significant influence can be seen by the change of the resistivity for subsequent post-annealing steps of 100°C and 140°C in table 5.2.

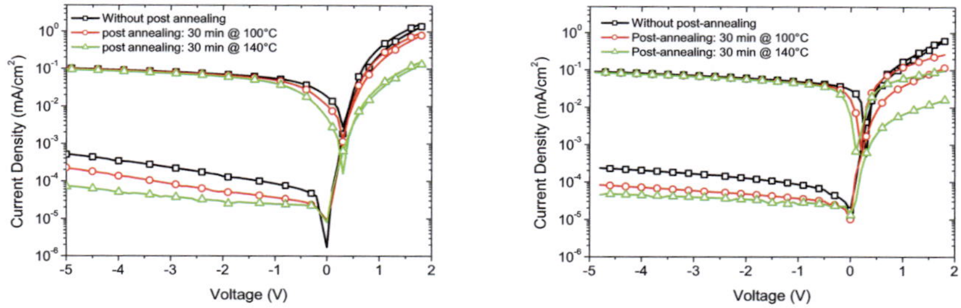

(a) Spray Coated

(b) Doctor-bladed

Figure 5.5.: Post annealing effect on spray coated (**a**) and doctor-bladed devices (**b**). Reducing dark and forward current densities for devices without post-annealing (black squares) by thermal treatment (post-annealing) with 100°C for 30 min (red circles) and with 140°C for 30 min (green triangles).

Deposition	**Dark Currents** (mA cm^2)			R_S (kΩ·cm^2)		
	non-annealed	100°C	140°C	non-annealed	100°C	140°C
Spray-coated	5.0×10^{-4}	2.4×10^{-4}	7.0×10^{-5}	0.6	1.0	6.0
Doctor-bladed	2.4×10^{-4}	8.5×10^{-5}	4.9×10^{-5}	1.5	7.9	83

Table 5.2.: Dark currents and serial resistances in dependence of thermal post-annealing

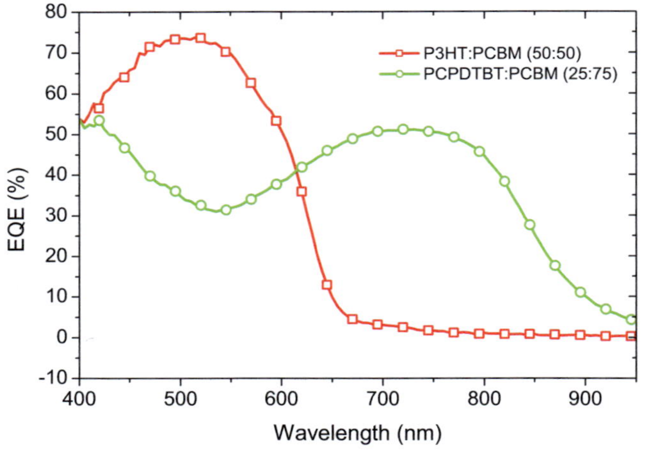

Figure 5.6.: Spectral response of pure P3HT:PCBM (red squares) and PCPDTBT:PCBM blends (green circles) at −5 V bias.

5.3. Tuning the Absorption Spectra

As discussed before PCPDTBT:PCBM blends in single devices are highly applicable for NIR detection. Another approach to engineer the spectral response from the visible up to 900 nm is a combination of the highly efficient P3HT:PCBM blend in the visible with the NIR sensitive PCPDTBT:PCBM composite. The spectral response for P3HT:PCBM and PCPDTBT:PCBM devices are shown in figure 5.6. In inorganic and organic semiconductor devices for photovoltaics, the most efficient devices are stacked configuration of different band gap semiconductors starting from the largest bandgap material and ending with the lowest bandgap material. A similar approach is chosen for stacking the PCPDTBT:PCBM films on P3HT:PCBM films (see figure 5.7(a)), where the low band gap material PCPDTBT: PCBM is deposited on top of the wide band gap P3HT:PCBM blend. Doctor-blading and spray-coating deposition are used to create a slowly merging or an abrupt junction of the two films, respectively. An intimate blend of PCPDTBT:P3HT:PCBM with varying contents is another possibility to modify the spectral response which is investigated within this section (see figure 5.7(b)). Engineering the spectral response in this range can be reached with conventional inorganic semiconductor only with huge efforts.

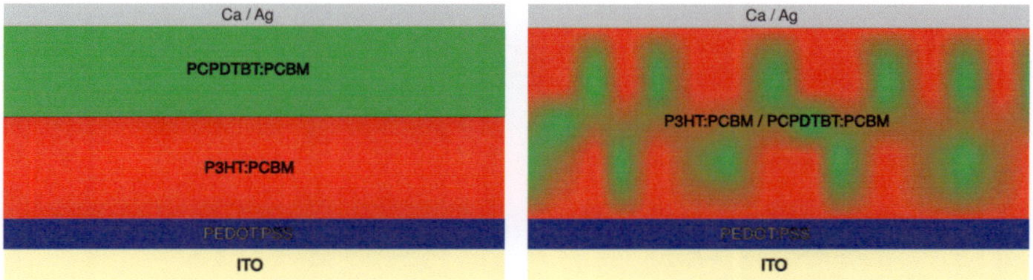

(a) Stacked device (b) Device of intimate films

Figure 5.7.: Combinations of P3HT:PCBM and PCPDTBT:PCBM blends for spectral absorption modifications: P3HT:PCBM and PCPDTBT:PCBM films stacked on each other (**a**) and blended together (**b**).

Stacked devices

P3HT:PCBM and PCPDTBT films were deposited subsequently by spray-coating and doctor-blading techniques. In figure 5.8(a) the I-V characteristics are shown and they differ only slightly in terms of photocurrents. With green illumination the single P3HT:PCBM film device exhibit slightly higher currents, whereas in the NIR the stacked devices show significant higher photoresponses compared to the P3HT:PCBM device. Spectral sensitivities shown in figure 5.8(b) for doctor-bladed stacked samples ($\sim 33\%$) are considerably higher than for spray-coated samples ($\sim 20\%$). The EQE of the stacked devices is a superposition of the P3HT:PCBM and PCPDTBT:PCBM films which is depicted as reference in figure 5.8(b).

Doctor-blading films on top of each other, dissolve partially the underlying layer again and allows a soft transition between the deposited layers whereas for spray-coating, the films are changed abruptly. Probably, the abrupt junction of the spray coated layers creates an interface with an increased density of traps causing higher recombination losses compared to doctor-bladed stacked films. Whereas doctor-bladed films allow a better charge transport/transfer from PCPDTBT:PCBM to P3HT:PCBM regions and vice versa due to interpenetrating domains of the films by dissolving the underlying layer. Comparably, simultaneous co-evaporation of a PPV derivative and C60 improve charge separation and transport by a good percolation of donor and acceptor [43].

Devices of intimate P3HT:PCBM / PCPDTBT:PCBM blends

Latter results on doctor-blading PCPDTBT:PCBM blends on P3HT:PCBM films indicate, that an intimate P3HT:PCPDTBT:PCBM composite might be more

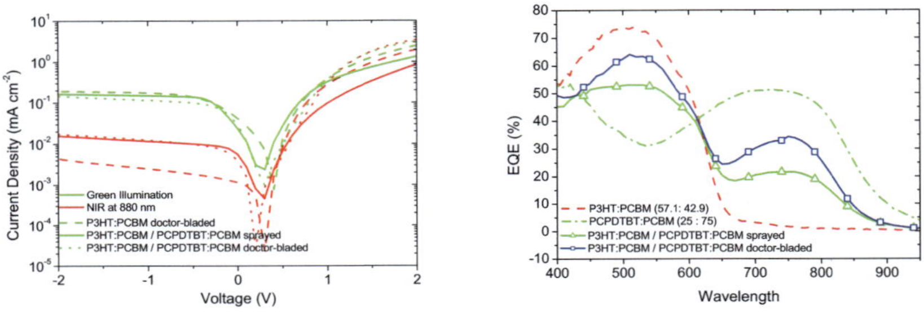

(a) Light currents

(b) EQE at -5 V

Figure 5.8.: Light currents and EQE of stacked devices. (**a**) I-V characteristics for spray-coated (solid) and doctor-bladed (dotted) stacked devices for green (green) and NIR illumination (red) at 880 nm.P3HT:PCBM doctor-bladed device is stated as reference (dashed). (**b**) EQE of spray-coated (green triangles) and doctor-bladed (blue squares) at -5 V bias. P3HT:PCBM and PCPDTBT:PCBM are shown as reference.

efficient. Another advantage is the better control on the contents of the material within the blend. Therefore, an adequate tool to engineer the spectral response is provided by modifying the content of materials. The solution for P3HT:PCBM (1:1) and PCPDTBT:PCBM (1:3) were solved separately 2 wt % in Chlorobenzene and blended together with ratios given in table 5.3. The different ratios of P3HT:PCBM (1:1) and PCPDTBT:PCBM (1:3) were chosen to keep the fullerene content high enough, because films with PCPDTBT need PCBM in excess for an efficient charge transport (see section 5.1) [112]. The devices were fabricated as described in chapter 3. In figure 5.9 the IV charac-

PCPDTBT:PCBM (1:3)	P3HT:PCBM (1:1)	P3HT (wt.%)	PCPDTBT (wt.%)	PCBM (wt.%)	EQE (%) at -5 V 515 nm	725 nm
1	10	45.5	2.3	52.3	56.6	6.0
1	2	33.3	8.3	58.3	51.0	26.4
1	1.5	30.0	10.0	60.0	46.9	39.1
1	1	25.0	12.5	62.5	45.2	35.0
1.5	1	20.0	15.0	65.0	41.7	42.4
3	1	12.5	18.8	68.8	32.5	47.1

Table 5.3.: Blending ratios, fractions and EQEs at 515 nm and 725 nm at a reverse bias of -5 V of each P3HT:PCBM and PCPDTBT:PCBM film.

teristics of the blends of P3HT:PCPDTBT:PCBM with varying contents are presented. A trend, that with growing PCPDTBT content the spectral properties of PCPDTBT are dominating is clearly observable. With increasing the PCPDTBT content from 2.3 wt% to 18.8 wt% the light current in the NIR is raised from 1.4×10^{-2} mAcm^{-2} to 7.3×10^{-2} mAcm^{-2} at -5 V while the light current in the green decreases from to 2.3×10^{-1} mAcm^{-2} to 1.3×10^{-1} mAcm^{-2} at -5 V. Note that the 12.5 wt% PCPDTBT blend does not correlate to the other results. For the experiment this device is considered as a statistical deviation (process fluctuations).

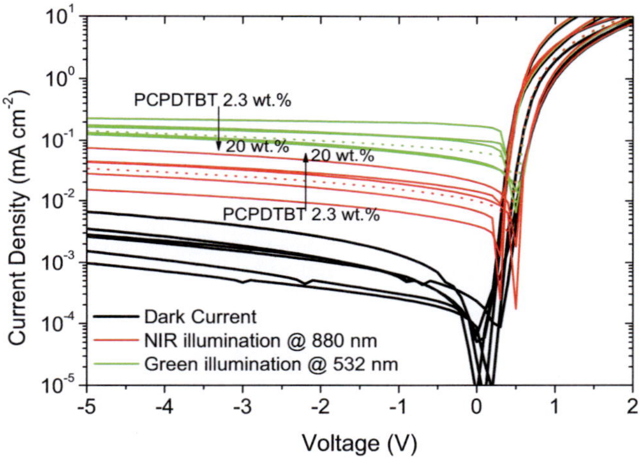

Figure 5.9.: I-V characteristics of different P3HT:PCPDTBT:PCBM contents in intimate films for dark (black line), green (green line) and NIR illumination at 880 nm (red line). The arrows indicate a growing PCPDTBT content, despite for the mismatching 12.5 wt% PCPDTBT blend (dotted line).

The strong modification is more clearly seen in the plot of the EQE in figure 5.10 in comparison to films with solely P3HT:PCBM or PCPDTBT:PCBM, respectively. The EQE at 515 nm drops from over 75% down to ~30%. On the other hand, in the NIR at 725 nm the EQE benefits from an increase of the PCPDTBT fraction: ~6% → ~51% (see figure 5.10(a)). EQE values at 515 nm and 725 nm for each blend are given in table 5.3. A divergent behavior is again observable for the 12.5 wt% PCPDTBT blend. When normalizing the EQE spectra to the maximum values, the 12.5 wt% PCPDTBT blend exhibits a flat level in the NIR spectral response in contrast to the peaks of the other ratios, what is another hint that this device is not representative. To summarize the superposition of the spectral response of the two blends is identified. To conclude, the engineering of the spectral response is demonstrated by blending P3HT:PCPDTBT:PCBM with different contents. The sensitivity, particular in the NIR region can be adjusted indeed by controlling the ratios of P3HT:PCPDTBT:PCBM.

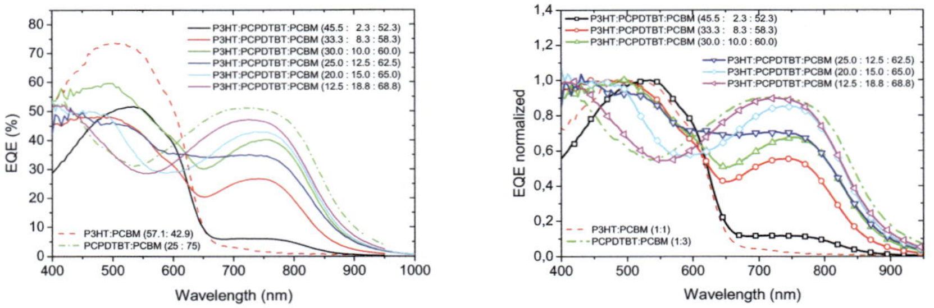

(a) EQE at -5 V

(b) Normalized spectral response

Figure 5.10.: Spectral Response of P3HT:PCPDTBT:PCBM devices with varying content concentrations. (**a**) With growing PCPDTBT content the NIR sensitivity increases significantly. (**b**) Normalized EQE spectra of intimate P3HT:PCPDTBT:PCBM blends. P3HT:PCBM (red dashed) and PCPDTBT:PCBM devices (green dash-dotted as reference.

In figure 5.9 a different field-dependence on the PCPDTBT content in the blend is distinguishable. In particular, the NIR response shows a strong field dependence with increasing PCPDTBT content for the light currents (figure 5.11(a)) and the NIR quantum efficiency (figure 5.11(b)), respectively. The more PCPDTBT in the P3HT:PCPDTBT:PCBM blend the more pronounced photoconductive NIR light currents can be observed (figure 5.11(a)) and a strong voltage dependence of the EQE at 725 nm (figure 5.11(b)) is observed as well. As pronounced photoconductivity is an attribute of PCPDTBT and not P3HT blends, the growing photoconductivity can be correlated to the increasing PCPDTBT content. For this reason P3HT:PCPDTBT:PCBM blends can be seen as a combination of the P3HT:PCBM and PCPDTBT:PCBM blends with reference to the corresponding weight fractions.

Bi-layer modeling for intimate multiple component blends

As discussed before, the blend behaves as a combination of independent P3HT: PCBM and PCPDTBT:PCBM devices, one can think of the reduction of the description to a simple bi-layer configuration. Simply speaking, the film shown in

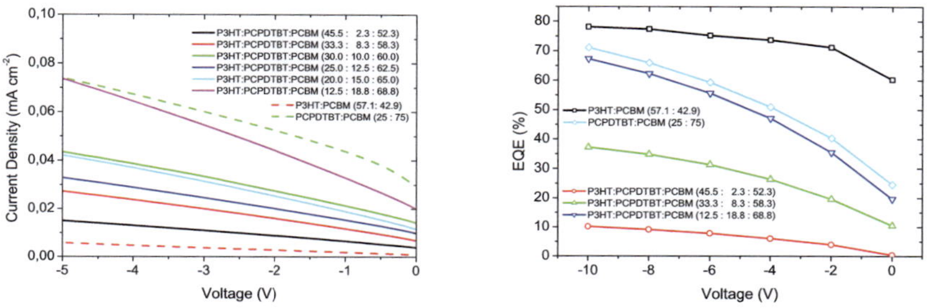

(a) Near-infrared light currents (b) EQE at 725 nm

Figure 5.11.: Voltage dependence of light currents (**a**) and EQE (**b**) according to the PCPDTBT wt.% content. Bias dependence of both, light currents and EQEs increase with PCPDTBT fractions.

figure 5.7(b) is transformed to a stacked structure (5.7(a)) considering the contents of the involved materials. Table 5.4 gives for each blend the total thickness and an effective thickness for the absorber material, which is calculated from the weight ratios of the composites. Since P3HT is widely investigated in terms of absorption, necessary absorption parameters can be found in literature [58]. For PCPDTBT a reliable absorption coefficient is determined by using the effective film thicknesses calculated from the PCPDTBT content in the films from table 5.4. As the absorption obeys Beer's law (see equation 2.42) and is proportional to $exp\,(-\alpha(\lambda)d)$, EQE values at 725 nm are used for fitting, because P3HT is not optical dense at this region. In figure 5.12(a) the measured EQE values are depicted as black squares and are fitted by

$$EQE = \eta_{ext}\,(1 - exp\,(-\alpha_{PCPDTBT}(\lambda)d_{PCPDTBT})) \tag{5.1}$$

with $d_{PCPDTBT}$ as the effective thickness of PCPDTBT in the film and η_{ext} taking into account the extraction efficiency for absorbed photons. By applying a least square fit, one obtains an absorption coefficient of $\alpha_{PCPDTBT}$ at $\lambda = 725$ nm of 2.47×10^5 cm^{-1} with $\eta_{ext} = 0.52$. The absorption coefficients are distributed spectrally for P3HT and PCPDTBT corresponding to their EQE spectra (see figure 5.12(b)). For PCPDTBT the determined absorption coefficient at 725 nm is used whereas for P3HT the coefficient was extracted from literature [58] with $\alpha_{P3HT}\,(\lambda = 525 \text{ nm}) = 1.8 \times 10^5$ cm^{-1}.

With the spectral absorption coefficients of each optical active component (PCBM is neglected as it is mainly optical active below 400 nm. However, PCBM is implicitly considered as the spectral response is included in the oBHJs

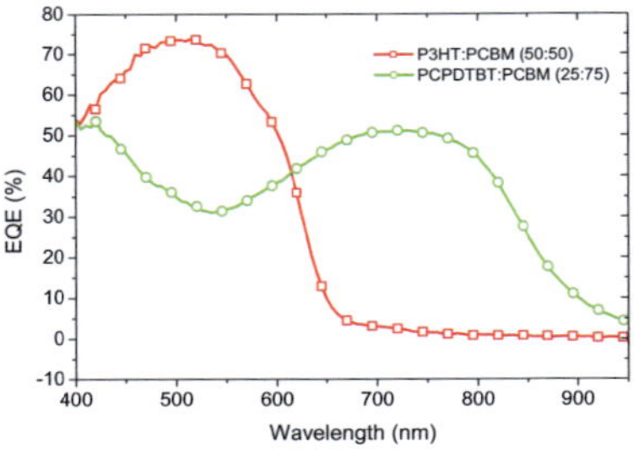

(a) Fitted spectral response at 725 nm

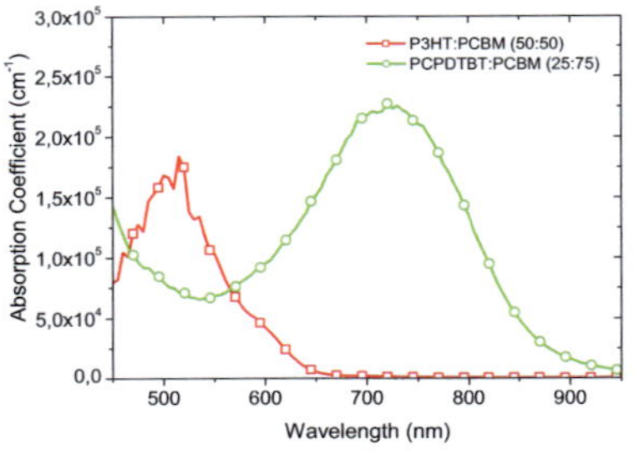

(b) Absorption Coefficients of P3HT:PCBM and PCPDTBT:PCBM

Figure 5.12.: Absorption properties of P3HT and PCPDTBT. (**a**) Exponential least square fit of near-infrared absorption at 725 nm for different blends taking the PCPDTBT content into account by an effective thickness (cf. table 5.4). (**b**) Spectral absorption coefficients for P3HT:PCBM blends derived from EQE spectra using peak absorption coefficients for P3HT from [58] and for PCPDTBT:PCBM blends with the extraction of the absorption coefficient from the fit in (a).

P3HT (wt.%)	PCPDTBT (wt.%)	PCBM (wt.%)	Film thickness (nm) total	d_{P3HT}	$d_{PCPDTBT}$	Fitting Parameters $\eta_{ext,P3HT}$	$\eta_{ext,PCPDTBT}$
45.5	2.3	52.3	430	196	10	0.580	0.296
33.3	8.3	58.3	450	150	37	0.503	0.456
30.0	10.0	60.0	412	124	41	0.641	0.670
25.0	12.5	62.5	391	98	49	0.559	0.565
20.0	15.0	65.0	448	90	67	0.440	0.547
12.5	18.8	68.8	600	75	111	0.329	0.499

Table 5.4.: Derived parameters for the different blends of PCPDTBT:P3HT:PCBM for film thicknesses and extraction efficiencies

of P3HT:PCBM and PCPDTBT:PCBM. Accordingly, the bi-layer model (figure 5.7(a)) can be applied now to fit the spectral response of the blends. First the absorption of the P3HT is considered for bi-layer devices as it is assumed that the larger bandgap material (here: P3HT) is the first layer succeeded by the low bandgap material (here: PCPDTBT) as second material which collects the remaining radiation. Note that the bi-layer model is an approximation and is not perfectly modeling the absorption an intimate film. This approach can be expressed by

$$\begin{aligned} EQE = \eta_{ext,P3HT} \cdot (1 - exp\,(-\alpha_{P3HT}(\lambda) d_{P3HT})) + \\ \eta_{ext,PCPDTBT} \cdot exp\,(-\alpha_{P3HT}(\lambda) d_{P3HT}) \cdot \\ (1 - exp\,(-\alpha_{PCPDTBT}(\lambda) d_{PCPDTBT})) \end{aligned} \tag{5.2}$$

with the η_{ext} as extraction efficiency for the absorbed photons, the absorption coefficients $\alpha(\lambda)$ and the effective thickness d for the corresponding material. In table 5.4 the fitting parameters are given for the best least square fit for each blend and the fits can be seen in figure 5.13 as red lines. For each blend a fit can be found which can approximately redraw the spectral response. Thus, the blends of P3HT:PCBM and PCPDTBT:PCBM can be transformed into bi-layer devices which can be described quite well by the superposition of two single devices considering the fractions of the materials. The model presented in this section takes the different contribution of each material into account by the effective thicknesses derived from the material fractions.

5.4. Transient Characteristics of NIR Polymer Diodes

The promising results show that PCPDTBT:PCBM photodiodes have a high potential for photodetection applications. Since many applications require suitable dynamic specifications, the photodiodes are investigated regarding their transient characteristics.

Noise analysis

In figure 5.14 the noise spectra of the photodiode and of the measurement setup are shown for voltages of -5 V. At a frequency of 170 Hz the photodiode has a noise current of 1.51×10^{-14} AHz$^{0.5}$, in contrast to the setup with 8.94×10^{-15} AHz$^{0.5}$. Noise currents are summed up squared $\hat{i}^2_{total} = \hat{i}^2_{DUT} + \hat{i}^2_{setup}$, thus the actual noise current of the photodiode is $\hat{i}_{DUT} =$

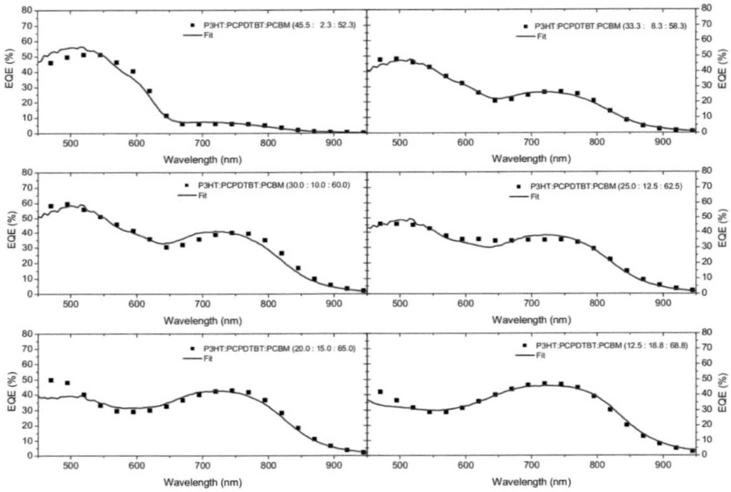

Figure 5.13.: Fits of EQE for different P3HT:PCPDTBT:PCBM ratios. The least square fit shows an adequate approximation for different ratios in the spectral region above 500 nm.

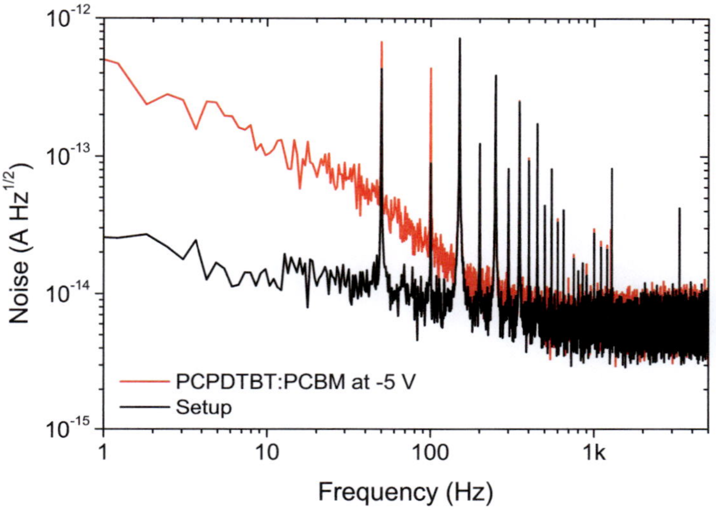

Figure 5.14.: Spectra of the noise current for a spray-coated PCPDTBT:PCBM device (red) and the measurement setup (black).

$\sqrt{\hat{i}^2_{total} - \hat{i}^2_{setup}} = 1.22 \times 10^{-14}$ AHz$^{0.5}$. With these values, one can derive the NEP and the D*:

$$\text{NEP} = \frac{\hat{i}_{DUT}}{EQE(\lambda = 725\text{ nm}) \cdot \sqrt{\Delta f}} = \frac{1.22 \times 10^{-14}}{0.501} = 2.44 \times 10^{-14}\ \text{WHz}^{0.5}$$

$$\text{D*} = \frac{\sqrt{A\Delta f}}{NEP} = 8.2 \times 10^{12}\ \text{cm (WHz}^{0.5}) = 8.2 \times 10^{12}\,\text{Jones}$$

Dynamic analysis

The dynamic properties of the photodiode are as well of importance, in particular, when applications need to process signals at high speed. Two different PCPDTBT:PCBM with different thicknesses of the active layer are investigated by pulsed photoexcitation at a wavelength of 660 nm. In figure 5.15 a device with an active layer of 550 nm shows the Bode plot of the amplitude and a signal at 10 kHz. One can recognize a low bass characteristic of the device with a -3 dB frequency at 130 kHz(5.15(a)). The fall and rise times can be extracted

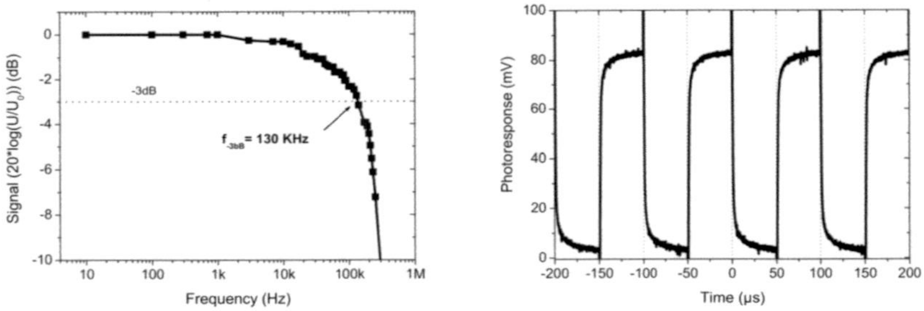

(a) Bode diagram of the amplitude

(b) Dynamic response at 10 kHz

Figure 5.15.: Dynamic behavior of a PCPDTBT:PCBM photodiode with an active layer thickness of 550 nm. (**a**) The Bode plot shows low-pass characteristics with a -3 dB frequency of 30 kHz. A signal train at 3 kHz with excitation at 660 nm is illustrated in (**b**).

from figure 5.15(b). Rise times of 4 µs and fall times of 5.1 µs are exhibited by the device. Most of the photogenerated carriers are created close to the electrode of the incident light. As ITO is the transparent and hole extracting electrode, contribution to the decay current from holes are neglected. Due to the close distance to the ITO and a higher mobility of 4×10^{-3} cm^2(Vs)$^{-1}$ of the holes [112], the decay current can be assigned to the electrons. Estimating the mobility from the transit time (approximated as fall time) with the given thickness and the voltage of -5 V one get reasonable values of $\sim 10^{-4}$ cm^2(Vs)$^{-1}$ (cf. ref. [112]).

For thicker device (800 nm) a slowdown of the photoresponse is assumed and proven in figure 5.16(a). The Bode diagram of the amplitude reveals a -3 dB, which is already reached at a frequency of 30 kHz. A signal train at 3 kHz is illustrated in figure 5.16(b), where the rise and fall times are determined to 33 µs and 36.5 µs. Deriving electron mobilities from the fall times with the same procedure as for the 550 nm sample gives values of 3.5×10^{-5} cm^2(Vs)$^{-1}$, which is almost lowered by a factor of 3 in relation to the 550 nm. Note that the calculation is to compare qualitatively the samples. Since the total optical absorption is reached already for roughly 250 nm, this effect might be assigned to a hindered charge transport in thicker films. In thinner films single domains for electron or hole conducting reach more probable the extraction electrode compared to thicker films. Thus, carriers (here: electrons) need to change domains to travel to the corresponding electrode, affecting the mobility.

In chapter 2.3.3 and in figure 2.31, a simple equivalent circuit for an organic photodiode has been introduced. The serial resistance and the capacitance of

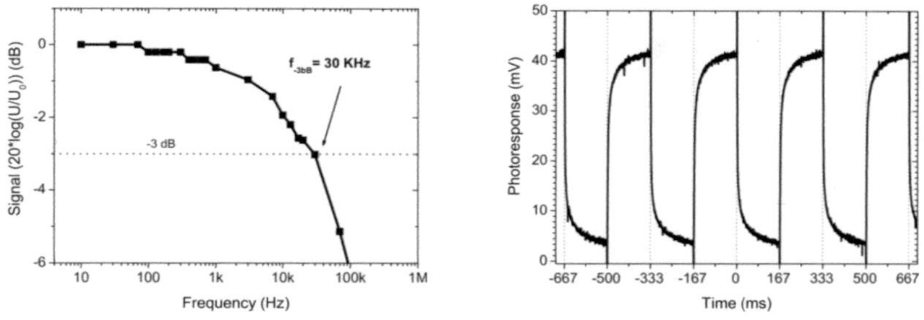

(a) Bode diagram of the amplitude

(b) Dynamic response at 3 kHz

Figure 5.16.: Dynamic behavior of a PCPDTBT:PCBM photodiode with an active layer thickness of 800 nm. (**a**) The Bode plot shows low-pass characteristics with a −3 dB frequency of 30 kHz. A signal at 3 kHz with excitation at 660 nm is illustrated in (**b**).

the photodiode are determining elements for the dynamic response of organic photodiodes (always assuming, that load circuits do not deteriorate dynamics). With lower mobilities, the resistivity for thicker films goes up resulting in a higher serial resistance. The serial resistance is therefore not exclusively proportional to the thickness, whereas the capacitance decreases reciprocal with the thickness. Higher transit can be explained by this relation resulting in a higher time constant.

6. Design and Optimization of PbS-Quantum Dot Sensitized Inorganic / Organic Hybrid Photodiodes

For detectors beyond 1 µm other materials than silicon need to be found since fairly no low-cost materials (e.g. organic polymers) are available for an efficient NIR detector as alternative. For wavelengths larger than 1 µm, expensive inorganic low band gap semiconductors such as In_xGa_yAs, PbS, etc are currently applied to state-of-the-art NIR detectors. For this reason, materials for cost-efficient processing from solution are of great interest. Besides challenging epitaxially grown devices such as quantum well infrared photodetectors, promising candidates are emerging materials from nanotechnology like colloidal QDs [55, 15, 56, 89] and soluble carbon nanotubes (CNTs) [92, 91]. Although these devices demonstrated attractive optical properties, stable and efficient devices for photodetection have been rarely reported. This chapter presents an approach for realizing IO-HPDs sensitized with QDs processed directly from solution. In the following, the relevant steps for optimizing the photodiodes are discussed corresponding to the specific layers. To recall the layout of an IO-HPDs the stack with all optional and mandatory layers for design purpose is illustrated schematically in figure 6.1.

6.1. Optimizing the Photoactive Layer

At the beginning there was the challenge to find suitable materials and compositions for near-infrared detection. NIR sensitivity beyond 1 µm was reported from several QDs made of II-VI semiconductors like lead sulfide (PbS) [105, 86], lead selenide (PbSe) [85] and mercury telluride (HgTe) [10]. Furthermore, colloidal QDs can cover a wide range of the UV-VIS (e.g. cadmium selenide [119]) and NIR spectrum [173]. PbS (bandgap: 0.37 eV [102]), PbSe (bandgap: 0.27 eV [101]) and HgTe (bandgap: 0.15 eV [103]) absorb as bulky materials deep into

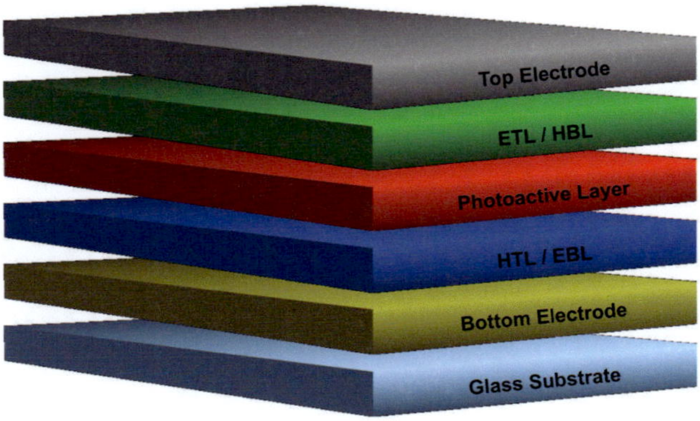

Figure 6.1.: Schematic illustration of a hybrid photodiode with solution processable QDs. All optional and required layers are shown. IO-HPDs are fabricated on a glass substrate with a bottom electrode. Hole transport layer (HTL) / electron blocking layer (EBL), photoactive layer from an inorganic-organic blend, electron transport layer (ETL) / hole blocking layer (HBL) and the top electrode are processed subsequently on top of a glass or backplane substrate.

the NIR/SWIR spectrum, so that these materials were selected for QDs of this work. However, in IO-HPDs no extension of the NIR sensitivity was observed with embedding PbSe- and HgTe-QDs. Similar, at a very early stage of this work soluble CNTs were discarded, since processing difficulties could not be overcome to reproduce the results from ref. [92, 91] and did not meet the expectations. Accordingly, this work focuses on PbS-QDs to enlarge the spectral response of IO-HPDs.

6.1.1. Composition of the Photoactive Layer

To begin with, known problems from colloidal QD-detectors exhibiting either poor charge transport properties [93, 56] or short lifetimes at maximum of several weeks [52, 85] need to be considered. Introducing inorganic-organic interfaces to hole and electron conducting organic semiconductors in the photoactive layer gives the possibility to use QDs as photosensitizers, while organic semiconductors facilitate the charge transport in the film. With this approach, the QDs can be used solely for donating photogenerated charge carriers to the organic semiconductors and can keep a protective ligand shell of OA. Photosensitization of polymer films by embedding PbS-QDs was already demonstrated by ref. [105, 189]. An advantage of PbS-QDs is that the synthesis is highly reproducible and scalable in an industrial range [55] compared to other colloidal QDs [90, 140]. Although charge transport properties of PbS-QDs and the usage of electron and hole conducting organic materials in oBHJ is well known, QDs have surprisingly never been casted together with a composition of an hole and electron conducting material. For this reason, at least the transport of one charge carrier type in the film is hampered due to insufficient charge transport properties of PbS-QDs. Thereinafter, the implementation of PbS-QDs into an hole conducting film of P3HT, an electron conducting film of PCBM and into an oBHJ of P3HT:PCBM is investigated for enhancement of the NIR sensitivity of hybrid photodiodes. With reference to figure 6.1, the layer stack of the samples consists of an ITO bottom and a Ca/Ag or Al top electrode, without any ETL/ or HBL and PEDOT:PSS layer as hole transport layer (HTL). The photoactive layers were deposited using doctor-blading technique (see 4.3 for a complete overview of the applied deposition technologies). A more detailed overview and the recent progress of inorganic-organic films for optoelectronic applications is given by ref. [56, 174, 88].

Another reason to choose PCBM and P3HT as host materials for the organic matrix to embed PbS-QDs is that the position of energy levels of the materials enables charge transfer from the Pbs-QDs to the organic semiconductors. In figure 6.2 the energy levels for colloidal PbS-QDs (first excitonic absorption peak at 1200 nm) in a P3HT:PCBM matrix are illustrated. The LUMO of -3.0 eV and the HOMO of -4.9 eV for P3HT are taken from ref. [155], while the energy levels of PCBM (HOMO: -4.3 eV and LUMO: -6.7 eV) were found in ref. [160]. Based on the energy levels of bulky PbS, the quantum effects for PbS-QDs are assumed to be symmetric for valence and conduction band as in PbS effective masses of holes and electrons are equal and the confinement of holes and electrons takes places simultaneously [116, 181], so that for the OA capped PbS-QDs a HOMO of -5.25 eV and a LUMO of -4.28 eV is obtained. Similar energy levels of OA capped PbS-QDs with an optical bandgap of 1 eV were reported from cycliovoltametry measurements in ref. [34]. With respect to the energy levels within in the composite, hole transfer from photoexcited PbS-QDs to P3HT should be possible as well as the electric field assisted electron transfer to PCBM. However, interactions of the PbS-QD interfaces to the organic materials are not taken into account for these considerations on energy levels [34].

In figure 6.3 I-V characteristics of devices with doctor-bladed films of dual (PbS-QDs:P3HT and PbS-QDs:PCBM) and ternary blends (PbS-QDs:P3HT:PCBM) are depicted. Note that for PbS-QD:PCBM the blend was drop casted on top of a doctor-bladed PbS-QD:PCBM layer. The PbS-QD weight content was kept constant at 66.7 % corresponding to 50 vol %. All devices were measured in dark and under illumination with NIR, green and white light. PbS-QD:P3HT devices show an asymmetric I-V curve with a pronounced photoconductive

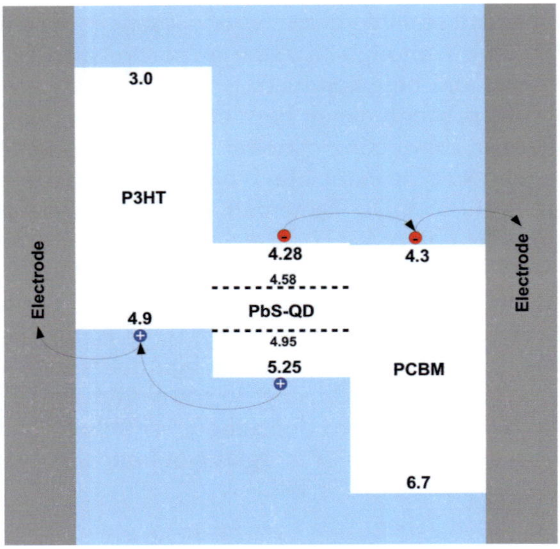

Figure 6.2.: Flat band energy diagram of PbS-QDs:P3HT:PCBM composite according to conventional semiconductor physics. For PbS-QDs with a first excitonic absorption peak at 1200 nm charge transfer to both materials appears to be possible. As indication the dashed lines are energy levels of valence and conduction band of bulk PbS, respectively.

behavior in forward as well as in reverse bias. In contrast, PbS-QD:PCBM films are dominated by low resistive properties showing little photoresponse in reverse direction. The ternary composition exhibits the closest photodiode characteristics. The light current is less field-dependent and shows a pronounced saturation of the light currents with minor photoconductivity. For each composition, the photoresponse for illumination can be observed in figure 6.3. Though a high photoresponse of the involved organic materials and of PbS-QDs in the visible is observed, significant light currents for NIR illumination for all three composites are found. Figure 6.4 shows the distribution of dark and light currents among the fabricated and working devices for each composition at a reverse bias of -5 V(see figure 6.4(a)) and a forward bias of 5 V (see figure 6.4(b)). Similar observation as in figure 6.3 can be found in current densities in figure 6.4. PbS-QD:PCBM films are very sensitive to conductivity and thus on process fluctuations. For this reason, the current densities in dark conditions and for different illuminations overlap strongly, so that light currents can differ from batch to batch. However, taking into account the mean values PbS-QD:PCBM composites exhibit correlated photoresponse for NIR illumination of 1×10^1 mAcm^{-2} compared to mean dark current densities of 8 mAcm^{-2} in reverse bias conditions. High currents in forward bias overlay light currents (compare figure 6.3, too) being not suitable for efficient photodetection in IO-

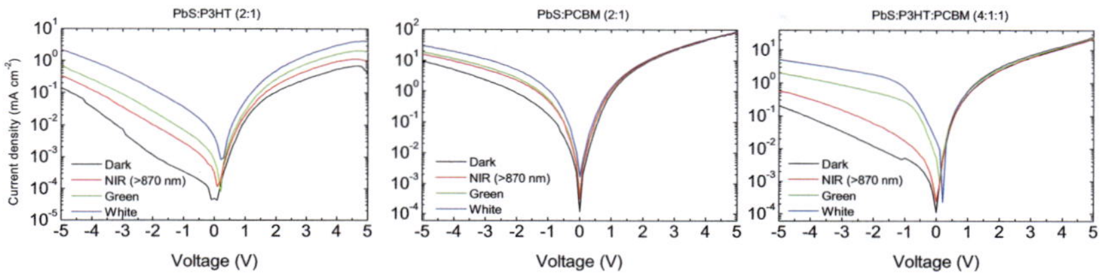

Figure 6.3.: I-V characteristics of PbS-QDs:P3HT, PbS-QDs:PCBM and PbS-QDs:P3HT:PCBM devices in dark conditions and with illumination at 532 nm (780 μW cm^{-2}), >870 nm and with white light of AM 1.5G (1100 μW cm^{-2}) . The ratio of the materials is given in weight percentage.

HPDs. PbS-QD:P3HT layers are characterized by much lower conductivity compared to PbS-QD:PCBM films resulting in lower dark and light currents.

The I-V measurements within this work give evidence to individual photosensitivities. To investigate the spectral response accurately, the IO-HPDs are further characterized by EQE measurements. In figure 6.5 the EQE for dual and ternary composite hybrid photodiodes at a reverse bias of -5 V are depicted. All of the components exhibit major photoactivity in the visible region. In dual component blends, the visible photoresponse is smaller as for the ternary blend, since the PbS-QDs need to operate as electron or hole transport material. PbS-QDs, however, are electrically insulated by an OA ligand shell, what is a restraint to charge transport. PbS-QD:P3HT:PCBM films are more sensitive to visible light. Since an oBHJ (P3HT:PCBM), which is also known from OPVs for efficient photoresponse in the visible, and PbS-QDs with an increasing spectral absorption starting from the first excitonic transition to the visible are implemented in the hybrid film. EQEs over 100 % can be described by photoconductive gain (see 2.1.4). In the NIR regime peak sensitivities can be found correlating well with the first excitonic transition of the PbS-QDs (see inset of figure 6.5). The response can be attributed solely to PbS-QDs as P3HT and PCBM do not absorb in the NIR. In figure 6.6(a) no absorption can be observed beyond 700 nm for P3HT and PCBM. In contrast, for PbS-QDs the first excitonic peak in absorption is clearly distinguishable (cf. figure 6.6(b)). Furthermore, composites of P3HT:PbS-QDs and PCMB:PbS-QDs reach EQEs around 1150 nm of 2.4 % and 1.6 %, respectively, ternary blends of PbS-QD:P3HT:PCBM achieve EQEs up to 7.8 % at -5 V. In contrast to dual composites, ternary composites demonstrate superior extraction of generated charges and therefore higher quantum efficiencies for hybrid photodiodes.

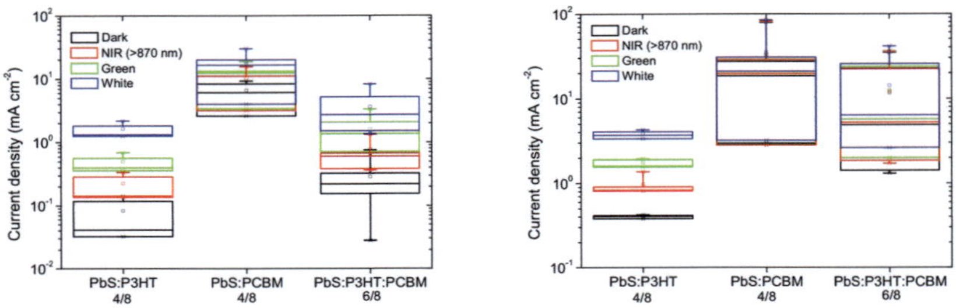

(a) Current densities at −5 V

(b) Current density at 5 V

Figure 6.4.: Distributions of dark and light currents for dual and ternary blends (**a**) at reverse bias of −5 V and (**b**) at forward bias of 5 V. The box plots gives the range from 25 % to 75 % and the whiskers from 5 % to 95 % of the values. The squares indicate mean values, whereas the crosses min and max values.The yield out of 8 devices on a single substrate is stated as well.

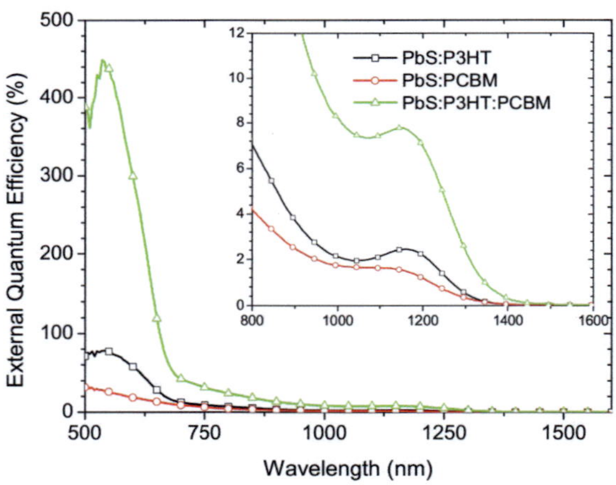

Figure 6.5.: Spectral response of devices with dual and ternary blends at a reverse bias of −5 V. The inset shows a zoom of the NIR regime related to the first excitonic absorption peak of PbS-QDs.

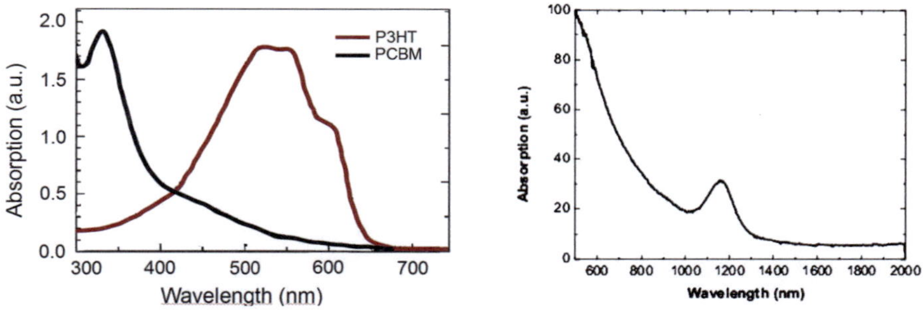

(a) P3HT and PCBM

(b) PbS-QD

Figure 6.6.: Absorption spectra of (**a**) P3HT, PCBM and (**b**) PbS-QDs films.

Since the implementation of electron and hole conductors should improve charge transport to extract both charge carriers more efficiently, the charge transport is analyzed by charge carrier mobility measurements with field-effect devices. Films with P3HyT:PCBM are characterized by an ambipolar transport with hole transport properties of P3HT and electron transport properties of PCBM in contrast for films with P3HT or PCBM only.The results and the conclusions from measurements on field effect mobilities are discussed in detail later in section 7.2.1. These measurements gives evidence to the importance to facilitate ambipolar charge transport for efficient hybrid photodetectors.

6.1.2. Concentration Dependence of PbS-QDs on Photodiode Performance

After identifying the appropriate compounds for NIR sensitivity, the composition of the blend is investigated and optimized. Four different weight contents of PbS-QD in an oBHJ of PCBM:P3HT ranging from 50 wt% to 80 wt% are investigated. Higher content of PbS-QD is preferred, since stronger absorption might increase the sensitivity of IO-HPDs in the NIR. Note that constraints in processing technologies and device physics like layer deposition and charge extraction did not allow applying thicker layers to optimize absorption. In figure 6.7 the different PbS-QD weight contents are characterized in terms of dark and light current densities (6.7(a)) as well as EQE at 550 nm, 880 nm and 1350 nm (6.7(b)) with an applied bias of -5 V. Note that for this experiment PbS-QDs with an first excitonic peak at 1350 nm were used which showed lower conductivity and sensitivity in comparison to PbS-QDs absorbing at 1200 nm to

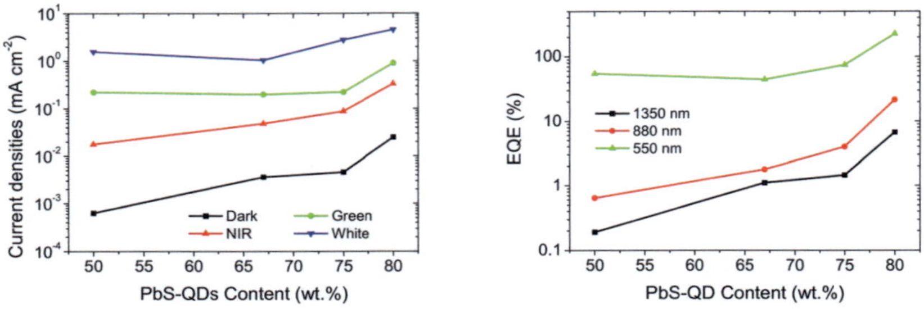

(a) I-V characteristics

(b) Spectral response

Figure 6.7.: I-V characteristics and spectral responses of IO-HPDs with increasing PbS-QD content. (**a**) Dark and light current densities illuminated by green,white and polychromatic light above 870 nm of IO-HPDs with increasing PbS-QD weight content in the film at a reverse bias of -5 V . (**b**) Spectral response of IO-HPDs with increasing PbS-QD weight content in the film at wavelengths of 550 nm, 880 nm and 1350 nm and a reverse bias of -5 V.

1250 nm. With increasing PbS-QD content the light currents rise at the expense of higher dark currents. Significant transitions according to changes in photoresponse and dark currents are from 50 wt% to 67 wt% and from 75 wt% to 80 wt% of PbS-QDs. Although the NIR response rises for films with 80 wt% of PbS-QDs, the dark currents increase dramatically from 6.3×10^{-4} mAcm^{-2} to 2.5×10^{-2} mAcm^{-2} by a factor of 38.7. However, the stabilization of the film by oBHJ vanishes with higher PbS-QD contents and more and more pinhole like defects occur. For concentrations of 67 wt% to 75 wt% the photoresponse and dark currents are comparable. As hybrid photodiodes for imaging applications should be characterized by low dark currents, a trade-off is suggested. Blends with 67 wt% and 75 wt% of PbS-QDs show suitable leakage currents and sensitivities. Whereas the strong increase of 80 wt% PbS-QD films in dark currents is not compensated by the higher photosensitivity for the current application. The variation in dark currents and EQE of films with different concentration of PbS-QDs is given in table 6.1.

PbS-QDs (wt%)	67 wt%	75 wt%	80 wt%
$I_{dark}/I_{dark,50\text{ wt\%}}$	4.7	6.3	38.7
$EQE_{dark}/EQE_{50\text{ wt\%}}$ at ($\lambda = 1350$ nm)	4.8	5.6	34.1

Table 6.1.: Relation with respect to 50 wt% PbS-QD films of dark current $I_{dark}/I_{dark,50\text{ wt\%}}$ and EQEs $EQE_{dark}/EQE_{50\text{ wt\%}}$ for ($\lambda = 1350$ nm) at an applied bias of -5 V.

6.2. Interlayers

In OLEDs and OPDs interlayers are key to modify electronic transport, especially injection and extraction properties of charge carriers. Thus, it is important to characterize and optimize the interface to improve the performance of hybrid devices by either enhancing extraction of photogenerated charge carriers or reducing leakages or even both. Interlayers are indicated as HTL / electron blocking layer (EBL) and ETL / EBL in figure 6.1.

Hole Blocking Layers

The most challenging interlayer of the hybrid photodiode is the HBL or ETL, respectively. HBLs need to be compatible to the stack processing, since HBLs are deposited on top of the active hybrid film. As a potential HBL additional 50 nm spray coating of PCBM was applied. Although from the position of HOMO at 6.7 eV and LUMO at 4.3 eV PCBM seems well suited as HBL, all experiments ended in short circuited devices. Thus, efforts were concentrated on EBLs / HTLs.

Hole Transporting Layer / Electron Blocking Layer

For interlayers working as HTL or EBL, the state-of-the-art interlayer of OPV and OPD devices is PEDOT:PSS and as an alternative a strongly hole blocking TFB was deposited on top of the ITO bottom electrode (anode). TFB is well suited as EBL in OPDs due to the position of the HOMO at 5.3 eV and the LUMO at 2.1 eV [21]. Thus, the high ohmic Poly[(9,9-dioctylfluorenyl-2,7-diyl)-co-(4,4'-(N-(4-sec-butylphenyl))diphenylamine)] (TFB) blocks very efficiently electrons in OPDs, from which hybrid devices possibly benefit in terms of leakage currents. However, TFB interlayer faced wetting properties with the hybrid blend resulting in inhomogeneous deposition of photoactive layer. To overcome this issue, TFB layers were either treated with O_2 plasma after deposition or an additional thin film of PEDOT:PSS was coated on top. The I-V characteristics with a thin topping PEDOT:PSS film and for O_2 plasma treated TFB are depicted in figure 6.8(a) and 6.8(b), respectively. For each experiment a reference with PEDOT:PSS interlayer was fabricated. Note, that the difference in the I-V characteristics of the reference devices is due to different PbS-QDs absorbing at 1220 nm and 1420 nm. For TFB/PEDOT:PSS devices, the dark currents could not be reduced and show a higher conductivity in reverse bias. In contrast, O_2 plasma treated TFB reduces significantly dark currents, but at the expense of poor response to NIR illumination ($<$870 nm).

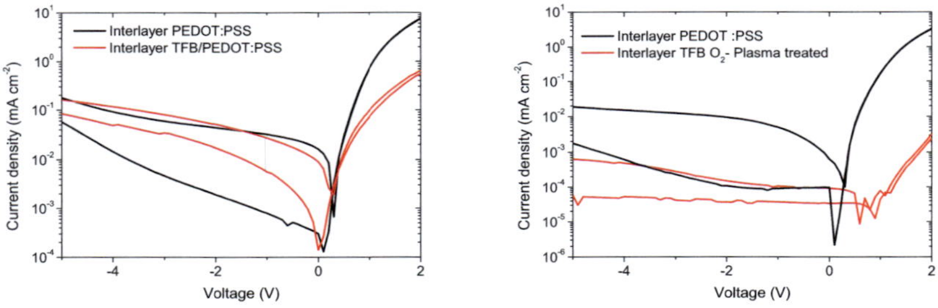

(a) TFB / PEDOT:PSS

(b) TFB O_2-Plasma treated

Figure 6.8.: I-V characteristics of hybrid photodiodes with TFB as interlayer under dark conditions and illuminated with polychromatic light beyond 870 nm of an AM 1.5G solar simulator. (**a**) TFB interlayer (30 nm) is sandwiched between ITO anode and PEDOT:PSS film. Reference with 1220 nm PbS-QDs. (**b**) Plain TFB interlayer with O_2 plasma treatment to increase wetting properties with a 1420 nm PbS-QD reference.

Spectral analysis confirmed the observations with I-V measurements of devices with TFB/PEDOT:PSS and O_2 plasma treated TFB interlayers. In comparison to PEDOT:PSS interlayers, the peak sensitivity of PbS-QDs diminishes significantly for TFB/PEDOT:PSS configuration or even disappears almost completely in spectral measurements for O_2-Plasma treated TFB device with PbS-QD absorbing at 1420 nm. Although dark currents and photoconductivity of the detector stack can be significantly influenced by TFB layers, those interlayer are not an alternative due to the poor observed sensitivities. In contrast to theory, where HOMO and LUMO levels of TFB layers promise an increase in performance, processing steps and interface interactions seem to have a strong negative effect on charge extraction.

The discussed results reveal, that for efficient hybrid photodiodes PEDOT:PSS interlayers are beneficial. Accordingly, a more detailed analysis is performed on the influence of PEDOT:PSS. PEDOT:PSS consists of two components: Poly-(3,4-ethylenedioxythiophene) (PEDOT) and Poly(styrenesulfonate) (PSS). In the following, the impact of self-assembled layers of PSS on top of the ITO anode on hybrid devices is investigated. ITO substrates were dipped for 30 s in a PSS solution of 0.5 wt %, 20 min and 30 min in a 0.1 wt % solution. Note that for dipping in 0.5 wt % PSS solution the substrates were washed with deionized water. In figure 6.10(a) with increasing the thickness of the PSS layer, the leakage currents of the devices can be reduced, but with a film thickness of 18 nm the diode properties already start disappearing. The reference device with PEDOT:PSS shows a superior pronounced diode behavior. Even worse

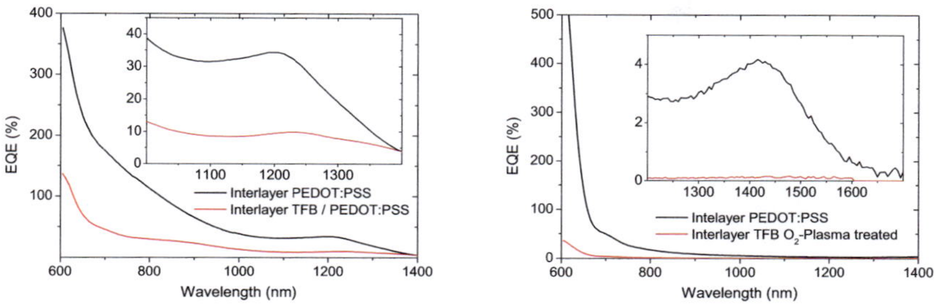

(a) TFB / PEDOT:PSS

(b) TFB O_2-Plasma treated

Figure 6.9.: Spectral response of hybrid photodiodes with either (**a**) TFB / PEDOT:PSS or (**b**) O_2 plasma treated TFB interlayer with a corresponding PEDOT:PSS reference. TFB interlayer is sandwiched between ITO anode and PEDOT:PSS film. Reference with 1220 nm PbS-QDs.

is the spectral response in the NIR where no response could be resolved for devices with self-assembled PSS layers (see figure 6.10(b)). Concerning the yield of fabricated devices, self-assembled PSS layers were poor in contrast to PEDOT:PSS devices. PEDOT:PSS interlayers are modifying more adequately the ITO anode interface for hole transport and for electron blocking from the photoactive layer as well as a better processing compatibility towards hybrid films compared to the investigated alternatives is given. Note that interlayers could not be skipped due to wetting properties of the hybrid films on ITO coated glass substrates.

To conclude, suitable interlayers to ensure smooth processing and to facilitate charge carrier extraction are essential for IO-HPDs.

6.3. Inorganic-Organic / Metal Interfaces

Organic films are strongly dependent on interfaces, especially to metals [12]. The built-in potential as well as the open circuit voltage of oBHJ devices are theoretically limited by the difference between donor HOMO and acceptor LUMO [11, 145]. However, oBHJ devices show also a strong dependence on the applied metal contact [109]. Work functions of the top metal electrodes and LUMOs of semiconducting organic materials define contact properties and therefore injection and extraction properties at the interface. For hybrid films

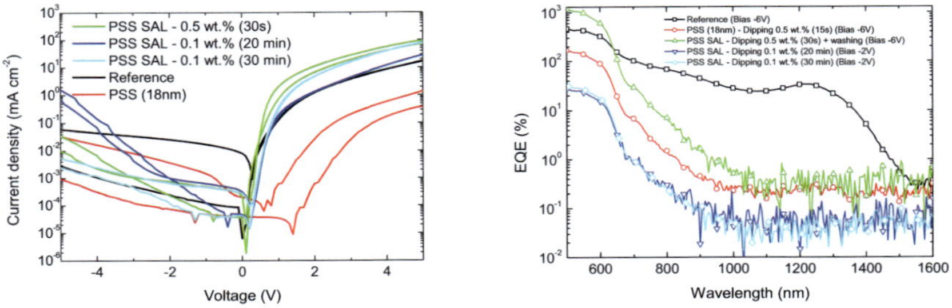

(a) I-V characteristics (b) Spectral response

Figure 6.10.: I-V characteristics (**a**) and EQEs (**b**) of devices with self-assembled layers of PSS modifying the ITO anode. The interlayer of the reference device is PEDOT:PSS.

an influence of work functions of metal contacts should be found correspondingly. Metal electrodes of LiF/Al ($\Phi_{LiF/Al} = 2.5$ eV), Ca/Ag ($\Phi_{Ca/Ag} = 2.9$ eV), Al ($\Phi_{Al} = 4.2$ eV) and Ag ($\Phi_{Ag} = 4.7$ eV) are investigated in the following.

In figure 6.11(a) I-V characteristics in the dark and under illumination currents and light currents with NIR-light are depicted. Ag and LiF/Al contacts exhibit large leakage currents and NIR light currents are hardly distinguishable. In contrast, Al and Ca/Ag top electrodes are showing distinctive diode behavior with a significant photoresponse. Al contacts give the best result for dark currents which becomes also evident with the on-off ratio (see figure 6.11(b)). The on-current at 2 V and the off-current at -2 V are put into relation and for Al and Ca/Ag top contacts a ratio of 7740 and 3900 is obtained whereas LiF/Al and Ag have poor on-off ratios of 40 and 213, respectively.

Highly rectifying contacts show also superior performance concerning the spectral response of the devices (see figure 6.12(a) for the visible and 6.12(b) for the NIR), since Al top contacts have EQEs of about 4 % at 1460 nm followed by 2.5 % at 1440 nm for Ca/Ag, 2.2 % at 1420 nm for Ag and 1.7 % at 1500 nm for Ag. For the shift of the peak in EQE no reason is found. High EQEs of Ag exceeding 200 % in the visible is due to ohmic contact of Ag to the oBHJ resulting in a strong photoconductive gain. However, the devices with superior diode characteristics lead to more efficient and sensitive IO-HPDs in the NIR. Figure 6.13 summarizes the findings that with Al top contacts the best results concerning dark currents and NIR photoresponse are obtained.

From a theoretical point of view, intuitively contacts with the lowest work functions should have a higher rectification, because of lower injection barriers for electrons and larger barriers in reverse bias for holes. This correlation is

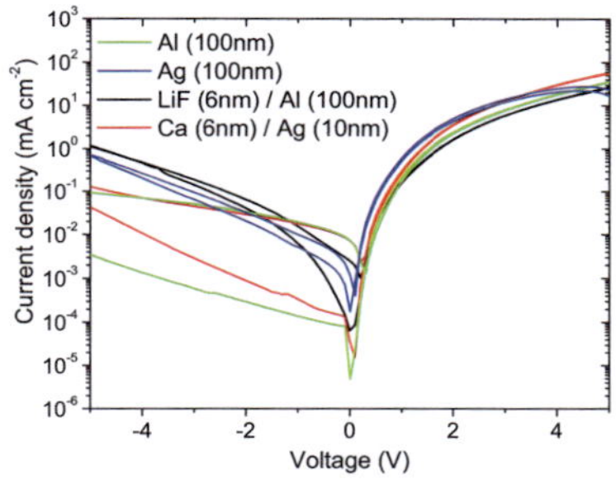

(a) I-V characteristics

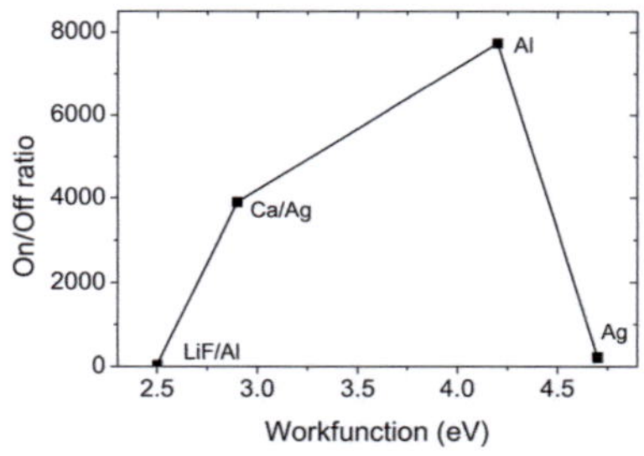

(b) On-off ratio

Figure 6.11.: (**a**) I-V characteristics of IO-HPDs with top contacts of different work functions in dark and illuminated with polychromatic light beyond 870 nm. (**b**) On-off ratio at ± 2 V related to the work functions of the top contacts.

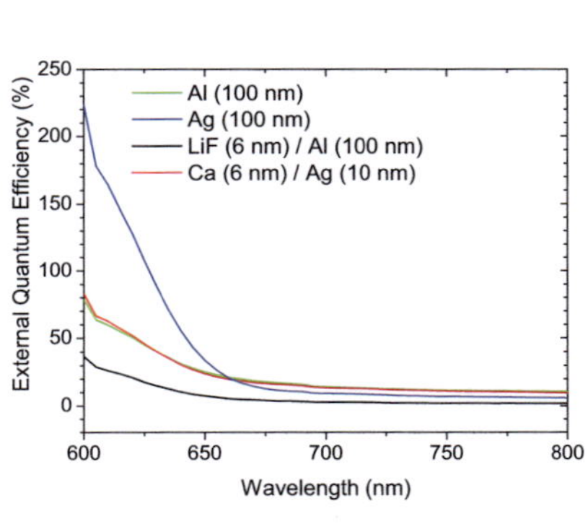

(a) Visible spectra

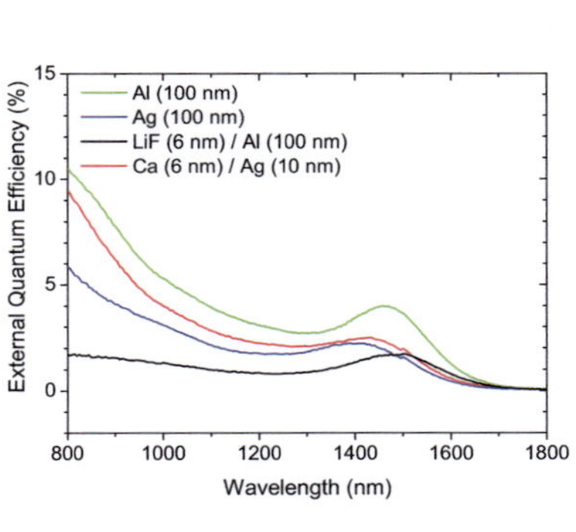

(b) NIR spectra

Figure 6.12.: Spectral sensitivity of hybrid devices with different top contacts in (**a**) the visible and (**b**) NIR.

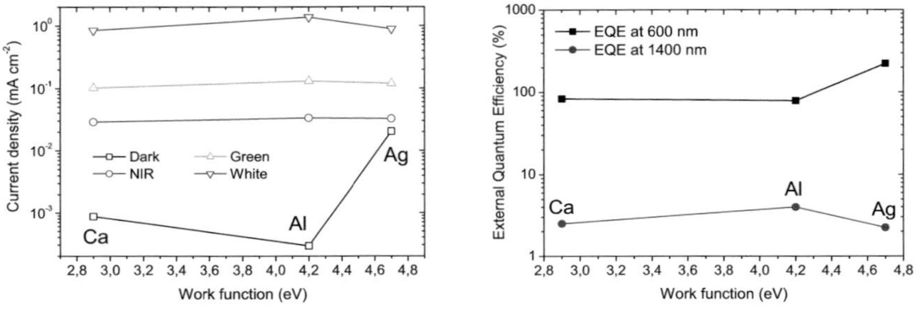

(a) Dark and light currents

(b) EQEs

Figure 6.13.: (**a**) Dark and light currents at a reverse bias of -5 V with relation to work functions of contact metals and (**b**) their efficiency for monochromatic illumination at 600 nm and 1400 nm with -5 V reverse bias.

apparently not applicable to hybrid films. Charge extraction properties without inducing leakage is achieved by Al contacts for PbS-QD:P3HT:PCBM films. Furthermore, the work function of Al is the closest to LUMO level of PbS-QDs and PCBM which should give an ohmic contact to these components (cf. figure 6.2). For smaller work functions of LiF/Al and Ca/Ag band bending occurs at the interface and electrons are transferred at the interface to the hybrid film. This incident with additional introduced states and traps from PbS-QDs might affect the extraction and rejection of charge carriers at the metal contact interface. However, no deeper investigations on interface physics were possible, thus the assumptions in this work rely on ref. [109, 12, 11].

6.4. Deposition Technologies

Solution processable organic photovoltaic and detector devices are very sensitive to morphology. A penetrated network of donor and acceptor is essential for organic devices and can be controlled by the deposition techniques and properties. Hence, hybrid films are as well dependent on deposition technologies. Spin-coating and drop casting of hybrid films were found to be unsuited. With spin-coating sufficient thick films could not be obtained and drop casting is not applicable for the focus on potential industrial processes. Nevertheless, doctor-blading and spray coating are well known technologies from organic photodetectors [176] and were adapted to hybrid films. So far results with

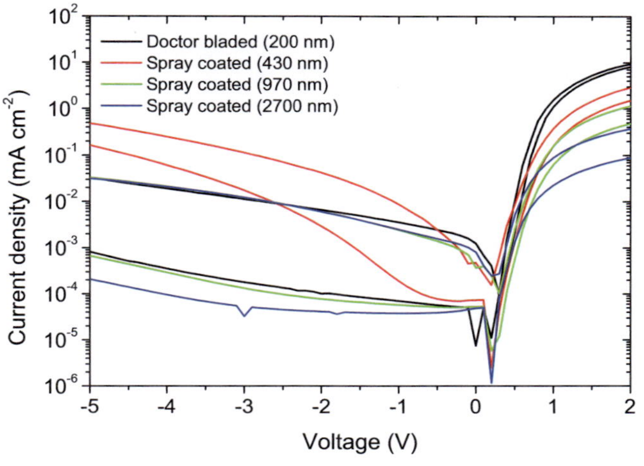

Figure 6.14.: I-V characteristics of spray-coated IO-HPDs of different thicknesses under dark conditions and illuminated with polychromatic light beyond 870 nm. A doctor-bladed sample is given as benchmark.

doctor bladed films up to 200 nm only have been presented, but especially to enhance the NIR absorption thick spray coated devices are favorable.

In figure 6.14 I-V characteristics of spray coated hybrid films of 430 nm, 970 nm and 2700 nm along with a doctor-bladed reference is presented. I-V performances of spray coated devices match the doctor bladed reference for thicker films than 970 nm. Spray coated films exhibit a high surface roughness and thus thicker films are needed to avoid pinholes (evident in the 470 nm spray coated device). The spray-coated device with a diode thickness of 970 nm is comparable to the doctor bladed reference with regards to dark and NIR light currents. Even lower dark currents are observed for the 2700 nm spray coated hybrid photodiode. Looking at figure 6.15 the results of I-V measurements are not in accordance with the NIR-EQEs of spray coated devices. The peak efficiencies for the doctor bladed device of 2.7 % at 1160 nm exceed significantly the spray coated diodes with 1.4 % and 0.6 % respectively (see figure 6.15). The matching light currents in the I-V characteristics can be attributed to the fact that spray-coated devices exhibit higher EQEs for wavelengths below 950 nm. Taking the electric field across the active layer into account a similar progression of the EQE at 1160 nm at low electric fields for all devices is observed, but at electric fields larger than 20 Vμm^{-1} the spray-coated devices either faces an irreversible breakdown or the EQE of the reference cannot be reached.

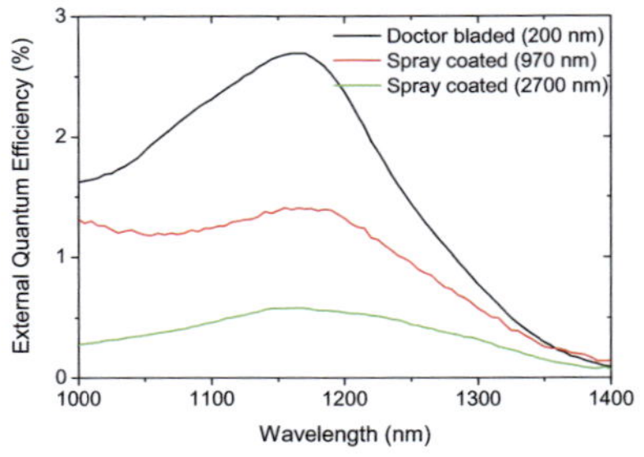

(a) Spectral response

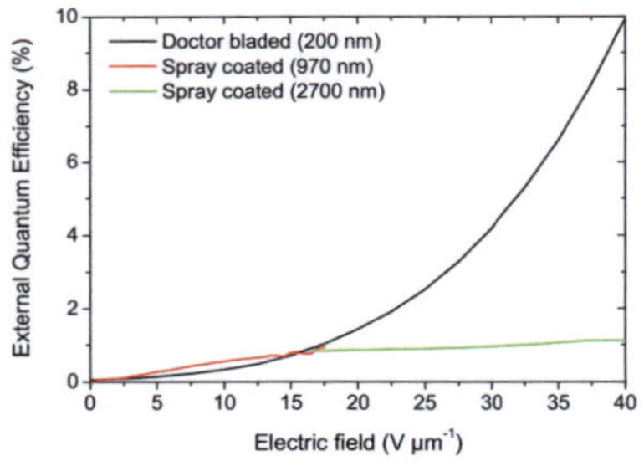

(b) Field-dependent EQE

Figure 6.15.: (**a**) EQEs for a 970 nm and a 2700 nm spray-coated device at −10 V with reference to a 200 nm doctor-bladed device at −5 V. (**b**) EQE according to the applied electric field across the active layer at a wavelength of 1160 nm.

Spray coated device showed promising results concerning dark and light currents, but also lack of efficiency for the NIR response as evidenced by the weak PbS-QD peak. Due to the high material consumption and restricted availability of PbS-QDs, no further optimization could be performed on spray-coated devices.

6.5. Influence of PbS-QD Properties on IO-HPDs

Electro-optical properties of PbS-QDs are influenced mainly by the diameter and the ligand shell. The first excitonic absorption peak is defined by the diameters of PbS-QDs. Energetic barriers and therefore charge carrier transfer and transport from and to the PbS-QDs can be modified by the ligand shells.

6.5.1. Variation of the PbS-QD Diameter

Within this chapter results with PbS-QDs absorbing at different wavelengths are presented. Figure 6.16(a) illustrates the spectral tunability with respect to the QD diameter. Increasing the diameter shifts the first excitonic absorption peak to longer NIR / SWIR wavelengths, so that sensitivities by modification of the diameter up to 1900 nm can be achieved. Although the diameter only

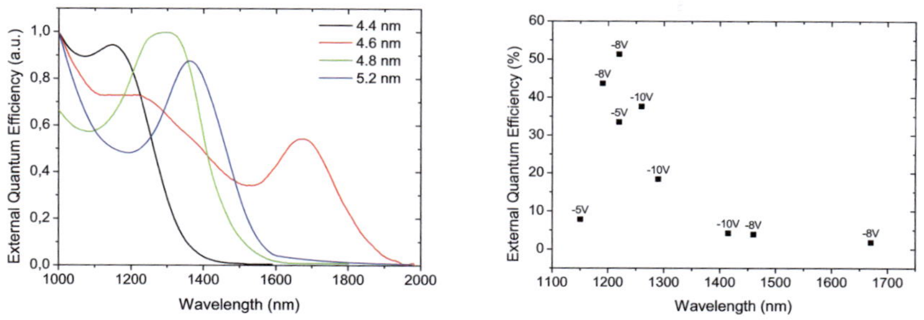

(a) Spectral response

(b) Field-dependent EQE

Figure 6.16.: (**a**) Tunability of the spectral response corresponding to QD-diameter. (**b**) Achieved EQEs for different QD-diameters. The applied bias during EQE measurements is given indicating a sweet spot for IO-HPDs with PbS-QDs at $\sim$ 1200 nm to 1250 nm.

changes the absorption peak, a significant impact on EQE performance of IO-HPDs can be attributed to PbS-QD diameters. A sweet spot for highly efficient hybrid diodes with PbS-QDs (around $\sim$ 5 nm) absorbing in the range of 1200 nm to 1300 nm is found. EQE values from 20 % to more than 50 % can be seen in figure 6.16(b). According to figure 6.2, for larger diameters the energy gaps of PbS-QDs become smaller and thus the transfer of photogenerated carriers is more and more hampered in particular for the transfer to PCBM by an occurring barrier. Therefore, hybrid devices with PbS-QDs of larger diameters can reasonably show less EQEs. For this exceptional behavior of PbS-QDs absorbing in the range of 1200 nm to 1300 nm a possible explanation might be a better resonant coupling of the first excitonic state to existing states in the oBHJ of P3HT:PCBM, so that those PbS-QDs have preferential transfer and transport properties. This would explain why PbS-QDs with smaller diameters have not shown those high efficiencies.

6.5.2. Blends with Various PbS-QD Diameters

Due to the tunability of the absorption of PbS-QDs, broadening of the spectral response might be obtained by embedding PbS-QDs of different diameters in a hybrid film. Actually, in figure 6.17 a superposition of the spectral response of the individual PbS-QD blends for the PbS-QD mixture can be observed, so that modification of the spectral sensitivity by implementing several different PbS-QDs could be demonstrated. However, the spectra could not be completely reconstructed by superposition from the EQE spectra of a hybrid device with

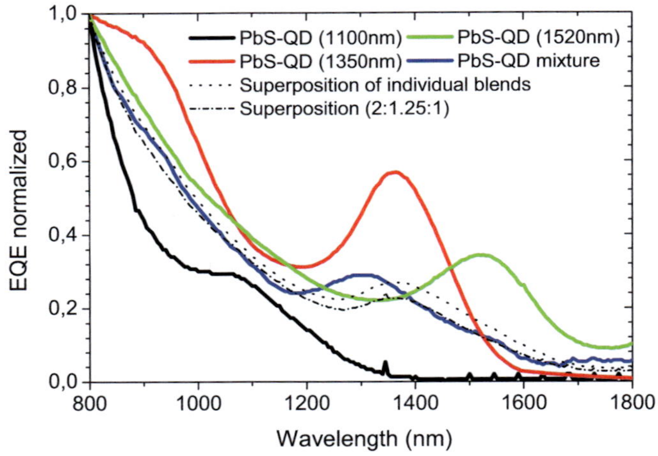

Figure 6.17.: Spectral response of IO-HPDs with QDs of three different diameters and their individual spectral response. A simple superposition and a weighted superposition is indicated to reconstruct the spectral response from the individual devices.

the individual PbS-QDs. The simple superposition gives better consistency with the absorption peak of the mixture while the weighted superposition of PbS-QD blends (1100 nm:1350 nm:1450 nm = 2:1:1.25) fits well to the tail. Although the spectral response was intentionally modified, the EQEs for the PbS-QD mixture ($\sim$0.15 %) could not reach the values for the individual PbS-QD devices by an order of magnitude. The interaction of different PbS-QDs in the film seems to interfere with the charge extraction and transport properties in comparison to hybrid devices with individual PbS-QDs, so that an introduction of additional trap states is assumed.

6.5.3. Ligand Shells

Oleic acid cappings stabilize PbS-QDs and at the same time electrically isolate the QDs [139, 31]. Among others, PbS-QDs can also be capped with butylamine ligands. The butylamine chain contains 4 carbon atoms in contrast to oleic acid (OA) with 18 carbon atoms. Therefore, butylamine ligands provide thinner enclosing barriers of PbS-QDs and higher conductivity for PbS-QD films [86]. Figure 6.18 gives the I-V characteristics (figure 6.18(a)) and the spectral sensitivity of IO-HPDs with embedded butylamine capped PbS-QDs (figure

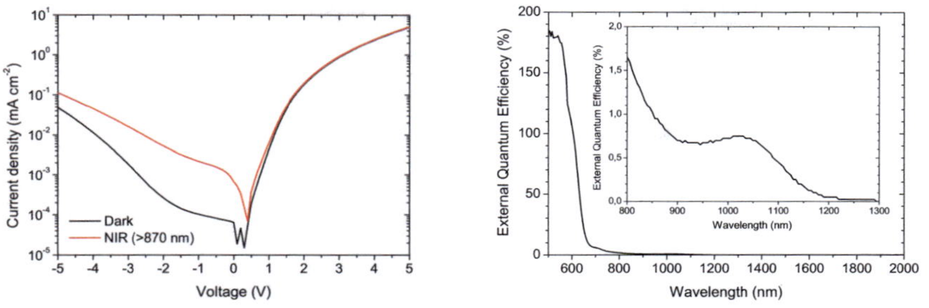

(a) I-V characteristics

(b) Spectral response

Figure 6.18.: (**a**) Dark and NIR light currents for IO-HPDs with butylamine capped PbS-QDs (**b**) EQE at a reverse bias of -8 V for hybrid butylamine capped PbS-QD devices.

6.18(b)). I-V characteristics show a rectification characteristic as for OA capped PbS-QDs, but with higher dark currents. The expected improvements on the transfer of photogenerated carriers could not be demonstrated which is apparent in poor EQE values of 0.75 % at 1020 nm (see figure 6.18(b). Since the properties of IO-HPDs could not be significantly improved with butylamine capping, OA is kept for ligand shell leading to longer lifetime of hybrid devices as discussed later in chapter 7.3.3.

6.5.4. Oxidation of PbS-QDs

Oxidation of PbS-QDs can modify the surface and therefore the barrier to transfer charge carriers from PbS-QDs[1]. PbS-QDs in chlorobenzene solution were exposed to highly saturated oxygen atmosphere for several hours. The surface of PbS-QDs is oxidized and the surface oxidation of the PbS-QDs is in a first approximation self-passivating. The PbS-QD itself is only affected on surface level and the oxidation process does not penetrate into the PbS-QD[1]. Oxidation was performed on parts of two independent PbS-QD batches dissolved in chlorobenzene. Afterwards, IO-HPDs were processed from the non-oxidized and oxidized part of the solutions. Different results are found due to oxidation as depicted in figure 6.19. Figure 6.19(a) reveals a significant improvement of the diode behavior. Oxidized PbS-QDs exhibit less field-dependent dark and light currents. In contrast only slightly lower dark currents can be observed in

[1]private communication with M. Bawendi, Department of Chemistry, Massachusetts Institute of Technology

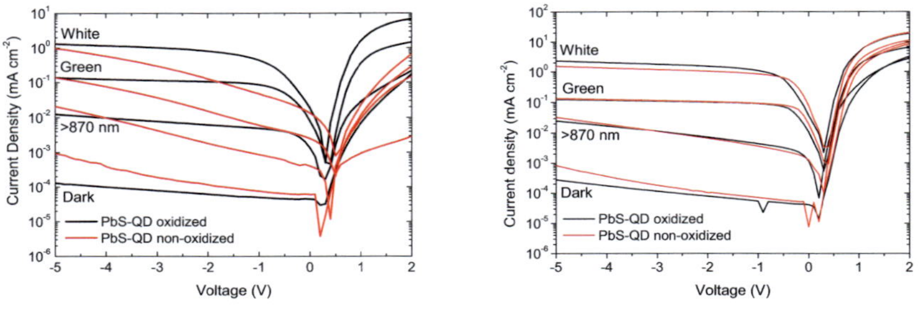

(a) First batch (b) Second batch

Figure 6.19.: Dark and light currents with green, white and NIR illumination measured for independent experiments on two individual PbS-QD batches. Each batch was partially oxidized.

figure 6.19(b). Apparently, the consequences of the oxidation differ for the two investigated batches and strongly depend on the batch properties.

In figure 6.20(a) and 6.20(b) varying spectral characteristics after oxidation of PbS-QDs were obtained. The first batch recovers after oxidation with a pronounced NIR absorption peak and with EQEs rising from 0.05% to 1.83% at 1170 nm. With a second QD batch oxidation leads to a decrease in EQE from 2.7% to 0.7% at 1170 nm. With oxdization the characteristic absorption spectra of PbS-QDs can be either restored or can be affected negatively. Correspondingly, similar observations have been made on the applied bias during EQE measurements (see figure 6.21). Whereas for the first batch less applied reverse bias is needed to extract photogenerated carriers, for the second batch higher voltages need to be applied to reach the same EQE.

Oxidation creates two observable effects on PbS-QDs. Either somehow a self-healing process recovers the PbS-QD properties and improves diode characteristics or PbS-QDs are negatively affected and electro-optical properties deteriorate. For the self-healing effect it is assumed that defects e.g. from missing ligands or impurities on the surface cured by oxidation. Oxygen atoms might occupy those states suppressing leakages (e.g. higher dark currents) and recombination centers by passivating the surface. In contrast, oxidation introduces defects to intact colloidal PbS-QDs as oxygen might interact with the ligands as well. Probably a self-healing effect occurs also for intact QDs when exposed long enough to oxygen.

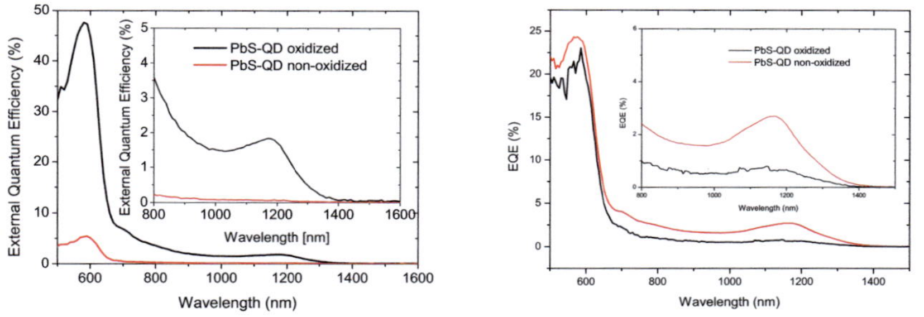

(a) First batch

(b) Second batch

Figure 6.20.: Spectral response obtained at independent experiments on two individual PbS-QD batch. Each batch was partially oxidized.

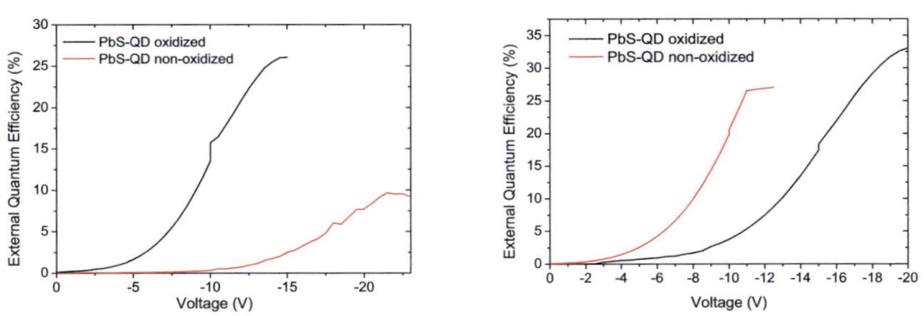

(a) First batch

(b) Second batch

Figure 6.21.: Voltage dependence of EQEs for two experiments on two individual PbS-QD batch. Each batch was partially oxidized.

Bottom Electrode

For reasons of completeness the bottom electrode is also an important subject for optimizations of IO-HPDs due to absorption losses, but is not further discussed within this work. In this work exclusively ITO anodes have been applied, but for ITO transmission decreases in the NIR from 80 % at 1000 nm to 30 % at 2000 nm (see figure 4.4).

6.6. The State of the Art IO-HPDs

As a conclusions, hybrid photodiodes based oBHJ show an optimum performance when following the design rules achieved in this chapter. A ternary blend of colloidal QDs with an electron and hole conductor for charge transfer and transport is advantageous. Doctor-bladed films of PbS-QDs in a P3HT:PCBM matrix showed the best results when combined with PEDOT:PSS interlayers and Al top electrodes. A sweet spot for the compatibility of PbS-QDs to the oBHJ was found for QD absorption peaks in the range of 1200 nm to 1300 nm. Therefore, the record device was made with a stack of ITO / PEDOT:PSS / PbS-QDs:P3HT:PCBM / Al. The achieved results are presented in figure 6.22. Polychromatic IR illumination increases the current density by almost 400 times from 8.8×10^{-5} mAcm^{-2} to 3.4×10^{-2} mAcm^{-2} at -2 V (see figure 6.22(a)). EQEs at 1220 nm of 16.5 % to more than 50 % according to the applied bias is obtained (see figure 6.22(b)). IO-HPDs are therefore highly suitable as NIR detectors.

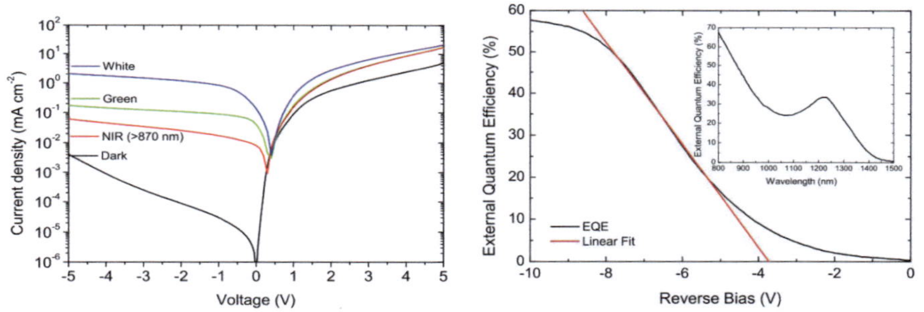

(a) I-V characteristics (b) Voltage dependency of EQE

Figure 6.22.: I-V characteristics of a hybrid device with an active area of 4 mm^2 in the dark (black line), with monochromatic illumination at 532 nm (780 $\mu W\,cm^{-2}$), with polychromatic light (> 870 nm) and with white light. (**b**) Voltage dependence of EQE with linear regime from -5 V to -8 V with an EQE of 51%. Inset: Spectral sensitivity of a photodiode with 4.5 nm PbS-QDs with a peak EQE of 16.5% at 1220 nm, biased at -5 V.

7. Characterization of Electronic, Optical and Structural Properties of Near-Infrared Inorganic/Organic Hybrid Photodiodes

In the previous chapter material, design and fabrication aspects without deeper analysis of optoelectronic and structural properties of hybrid photodetector devices were discussed. To complete the picture of IO-HPDs this chapter gives more insight into morphology, charge carrier transport and transfer, optical and electro-optical characteristics.

7.1. Structural Analysis by SEM and TEM Methods

Since morphology has a major impact on device characteristics of nanostructured organic and hybrid films, investigations with SEM and TEM methods on freshly processed devices were performed. For a detailed description of methodology for SEM and TEM see section 3.2.1 and 3.2.2, respectively. As the devices were not encapsulated after processing for the SEM and TEM analysis, only restricted I-V measurements under inert atmosphere were recorded. In chapter 6 dual composite and ternary composites were compared which showed superior photodetector properties. For SEM analysis the optimized ternary composite and a bilayer configuration of PbS-QDs:P3HT (66.7 wt%:33.3 wt%) were selected as they were the only devices with a significant NIR response of this processed batch. Dual composites were not investigated since the better results for ternary devices could be attributed to charge transport properties. Note that bilayer devices exhibit dramatically lower NIR response than hybrid detectors of ternary composites. Bilayer configurations (cf. figure 7.1) have a sharp transition between PbS-QD layers and P3HT while the PbS-QDs are distributed widely throughout the ternary film (cf. figure 7.2).

Following the principle of oBHJs [143, 185], charge transfer and transport is better facilitated in penetrated network of conducting and charge carrier donating materials as in separated films, especially for highly insulated OA capped PbS-QDs. In figure 7.1 two bilayer devices are shown: a doctor-bladed PbS-QD film annealed for 60 min at 200 °C with a second P3HT doctor-bladed layer (see figure 7.1(a)) and not annealed with a spray-coated P3HT layer on top (see figure 7.1(b)). The annealed PbS-QD film acts as a blocking layer as from I-V measurements very poor NIR response and very low dark currents in the range of 10^{-5} mAcm^{-2} at -5 V were obtained, although it is assumed that ligands are removed from PbS-QD surface for temperatures of 200°C. PbS-QDs are not dissolved from the P3HT solution during the subsequent doctor-blading process. The spray-coated P3HT layer partially dissolves the PbS-QDs again. Tiny PbS-QD bubbles above the QD film can be seen. Higher dark currents (10^{-2} mAcm^{-2} to 10^{-1} mAcm^{-2} at -5 V) and NIR response were observed. A possible reason for leakage was identified by paths through large PbS-QD agglomerations connecting top and bottom electrodes (see figure 7.1(b)).

(a) Doctor-bladed bilayer

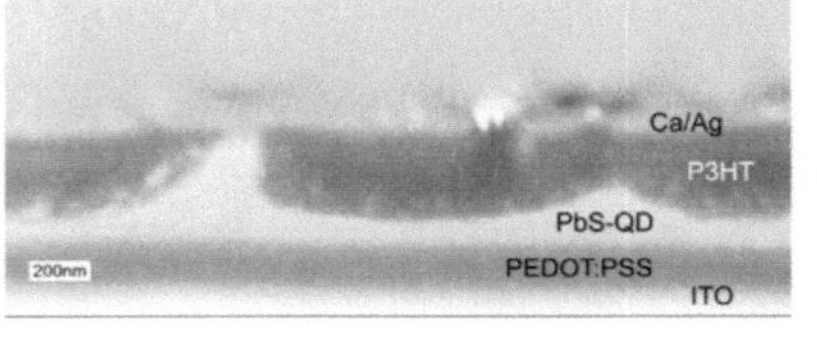

(b) Spray-coated bilayer

Figure 7.1.: SEM images of bilayer devices of PbS-QD:P3HT. (**a**) PbS-QD doctor-bladed and annealed for 60 min at 200 °C before doctor-blading of P3HT. (**b**) PbS-QD doctor-bladed before spray-coating the P3HT layer without annealing the PbS-QD film.

As already the spray-coated interlayer with partially dissolved PbS-QDs indicated a interpenetrating network being favorable for efficient charge transfer and extraction. Highly efficient IO-HPDs were fabricated from ternary composites of PbS-QD:P3HT:PCBM processed from solutions (see chapter 6). Figure 7.2 illustrates the PbS-QD distribution in an organic matrix of P3HT:PCBM for a doctor-bladed (see figure 7.2(a)) and a spray-coated film (see figure 7.2(b)). Doctor-bladed films exhibit a more homogeneous distribution of PbS-QDs in the oBHJ. In contrast, cloud like domains for PbS-QDs of roughly 200 nm in size in spray-coated films can be found. This can be a possible reason for the lower efficiencies of spray-coated devices as spray-coated films of ternary composites provide fewer surface for the organic/inorganic interface to transfer excited charges (cf. 6.4). Note that because of strong charging of the samples during SEM some images are blurred.

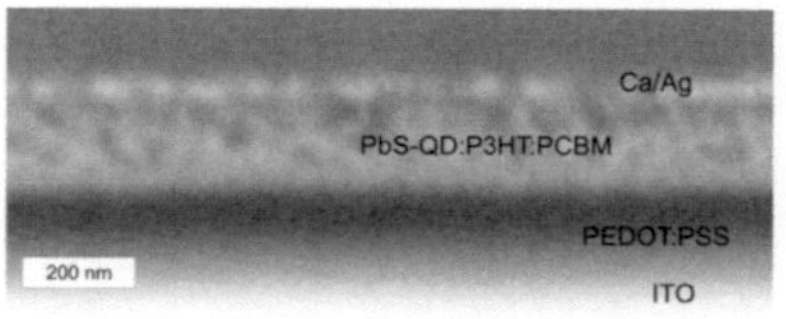

(a) Doctor-bladed

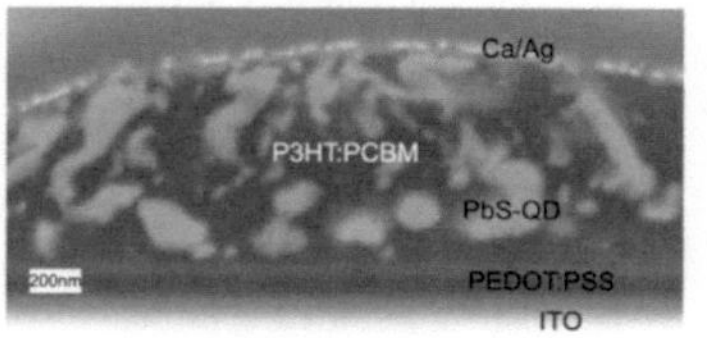

(b) Spray-coated

Figure 7.2.: SEM images of ternary composite devices with (**a**) a doctor-bladed and (**b**) a spray-coated PbS-QD:P3HT:PCBM film.

For a better spatial resolution TEM was performed on doctor-bladed ternary films. A 200 nm scale TEM image is depicted in figure 7.3(a) where the PbS-QDs can be distinguished from the organic as dark dots. The homogeneous distribution of PbS-QDs in the oBHJ is confirmed by TEM images. In figure 7.3(b) an ensemble of three PbS-QDs can be identified in the white circle by the crystalline lattice structure.

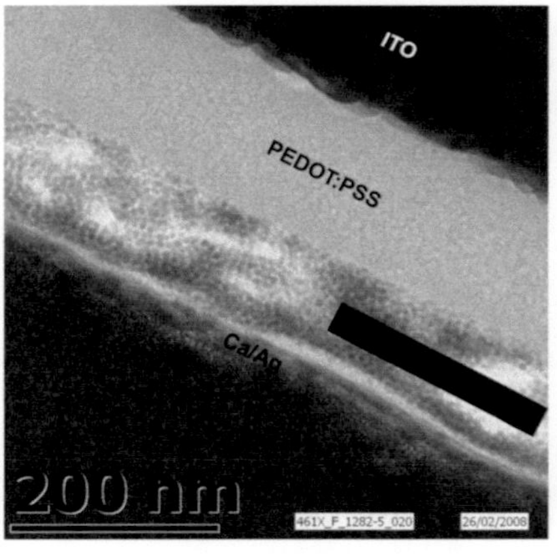

(a) 200 nm scale

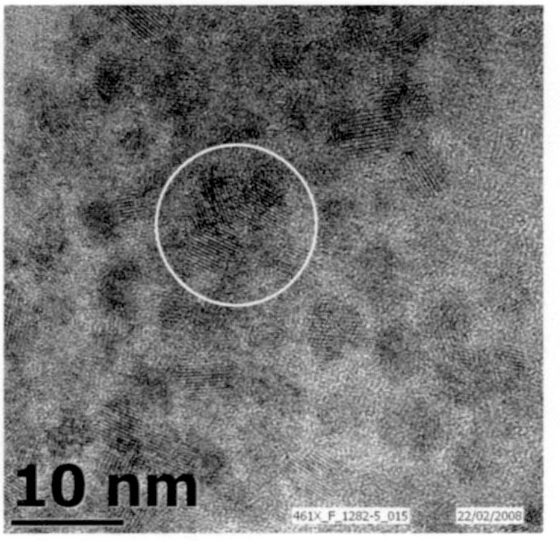

(b) 10 nm scale

Figure 7.3.: TEM images of a doctor-bladed PbS-QD:P3HT:PCBM film for (**a**) a 200 nm and (**b**) a 10 nm scale.

SEM and TEM analysis revealed that colloidal PbS-QDs can be embedded homogeneously into organic matrices and that highly penetrated networks of inorganic/organic composites can be obtained being favorable for efficient charge extraction. This corresponds to findings in OPVs and OPDs.

7.2. Charge Transport and Transfer Mechanisms in hybrid Films

Since experiments on different compositions of the active blend in hybrid photodiodes proved that it is essential for efficient near-infrared sensitivity to use both, a hole conductor (P3HT) and an electron conductor (PCBM) in the film (see section 6.1). Thus, the ternary blend is advantageous in terms of charge transfer and transport of photogenerated carriers. In this section, single and multiple composite blends are investigated concerning charge transfer mechanisms and charge carrier mobilities.

7.2.1. Mobilities in Hybrid Films

For mobility analysis (see 7.2.1 and 7.2.1), two common methods - mobility extraction from FET and CELIV - were used for a closer look on the mobilities in films and in the bulk of the hybrid photodiodes. Charge transport is essential to realize efficient photodiodes. Hence the photogenerated carriers need to be extracted at the electrodes.

Field Effect Mobilities

In OFETs either electrons or holes can be accumulated at the gate oxide with applying a gate voltage V_{GS}. Mobilities of electrons μ_e and holes μ_h can be studied and derived from e.g. the transfer characteristic of an OFET. Mobilities in the films should give information if any blend is better capable for charge transport. Findings in OPV revealed the relation between a balanced charge transport and the efficiency ([111, 112]). In table 7.1 all investigated compositions of the films as well as the weight ratio of the components and the obtained mobilities with the applied gate V_{GS} and drain voltage V_{DS} are given. Actually, besides the BHJ, only the ternary blend exhibits both, hole and electron mobility.

In figure 7.4(a) the transfer characteristics of PbS-QD:P3HT and PbS-QD:PCBM with reference to measurements on single component films are depicted. P3HT and PCBM show hole or electron conductivity, respectively, whereas PbS-QDs can be seen as non-conductive for any charge carrier. Note

Nr.	Composite	Weight ratio	μ_e $(cm^2(Vs)^{-1})$	μ_h $(cm^2(Vs)^{-1})$	V_{GS} (V)		V_{DS} (V)	
					μ_e	μ_h	μ_e	μ_h
1	P3HT	—	—	1.6×10^{-2}	—	-37.0	—	+14
2	PCBM	—	1.1×10^{-2}	—	+26.2	—	-18	—
3	PbS-QD	—	6.1×10^{-8}	—	+32.0	—	-12	—
4	PbS-QD:P3HT	2:1	—	2.6×10^{-3}	—	+37.0	—	+18
5	PbS-QD:PCBM	2:1	7.3×10^{-4}	—	29.7	—	+16	—
6	P3HT:PCBM	1:1	1.2×10^{-5}	1.5×10^{-2}	+29.7	-36.3	+18	+18
7	PbS-QD:P3HT:PCBM	4:1:1	4.6×10^{-5}	5.3×10^{-4}	+36.6	-36.9	+18	+18

Table 7.1.: Analyzed composites in OFETs and the measured mobilities with corresponding voltages V_{GS} and V_{DS}.

that the PbS-QDs are capped with long oleic acid ligands. For both films either a channel is formed for electrons or holes and a reduction due to PbS-QD addition is significant. A similar behavior can be observed for the ternary blend in figure 7.4(b). The ambipolar - electron and hole - mobility are affected by the QDs. In figure 7.5 mobilities are illustrated according to the film composition. Although 1.6×10^{-2} cm^2(Vs)$^{-1}$ as hole mobility for P3HT exceeds the mobility reported by [111] by almost two orders of magnitude, values in the same regime can be found in literature as well (see [158, 4]). As the surface was treated differently with oxygen plasma before deposition to [111], the effect is assigned to interactions at the interface of the oxide and the P3HT-film where the channel forms (higher degree of ordering). Electron mobilities of 1.1×10^{-2} cm^2(Vs)$^{-1}$ for PCBM correspond to reported values in literature [111]. The results from P3HT:PCBM films are inconsistent as the hole mobility is 1.5×10^{-2} cm^2(Vs)$^{-1}$ whereas the electron mobility is 1.2×10^{-5} cm^2(Vs)$^{-1}$. The discrepancy of the hole mobility being close to mobility in pristine P3HT to the dramatic reduction of the electron mobility can be explained by the morphology of the film. It is assumed that the channel region consists solely of P3HT, so that the properties are dominated by P3HT. Note that the authors of [111] used a method with which this effect can be considered, but this method was not applied to these measurements as the appropriate structures were not available. While pure PbS-QDs exhibit very small electron mobility of 2.6×10^{-8} cm^2(Vs)$^{-1}$, adding PCBM (2:1 in weight) as electron conductor improves the mobility to 7.3×10^{-4} cm^2(Vs)$^{-1}$. On the contrary for the PbS-QDs blended in P3HT (2:1 in weight as well), only hole conductivity is observed with a mobility of 2.6×10^{-3} cm^2(Vs)$^{-1}$. The ternary blend with a weight ratio of (4:1:1) results in an ambipolar transport with a hole mobility of 5.3×10^{-4} cm^2(Vs)$^{-1}$ and an electron mobility of 4.6×10^{-5} cm^2(Vs)$^{-1}$. Note that the applied method gives only an indication of the mobilities as field effect mobilities are strongly bound to interface properties of films and gate oxide.

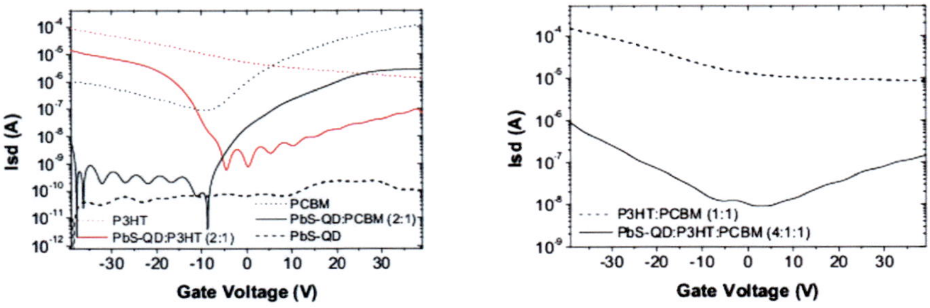

(a) Single and dual composites

(b) Ternary composites

Figure 7.4.: FET transfer characteristics. (**a**) Dual composites in comparison with single components show an expected behavior as the current in the FET is reduces significantly. (**b**) Analogous, the current for a standard P3HT:PCBM BHJ is decreased significantly with the addition of PbS-QDs.

The mobility results confirm the assumption that an organic matrix of both hole and electron conductor enables an ambipolar transport in the composite diode as a prerequisite to obtain high EQE values reported for the IO-HPDs. Thus, providing transport possibilities for holes and electrons is crucial to separate and withdraw photogenerated charge carriers from the QDs via percolation paths. In contrast, the EQE of P3HT or PCBM layers sensitized with PbS-QDs (67 wt%) show significantly lower values of 0.2% and 2.3% at -5 V, respectively. Whereas, IO-HPD based on the ternary blend yield high EQEs in the infrared at 1220 nm of 16.5% at -5 V and even 51% at -8 V (cf.6.1.1 and 6.6).

Ambipolar Bulk Mobilities by CELIV Method

Since mobility measurements on FET devices are mainly afflicted with interface and channel properties, CELIV measurements were performed to gain insight to charge transport in the bulk of PbS-QD:P3HT:PCBM devices. CELIV methods give information about the equilibrium charge carriers and do not distinguish between the charge carrier types. In figure 7.6 the obtained curves from the CELIV measurements in the dark (figure 7.6(a)) and from Photo-CELIV measurements at 1310 nm (figure 7.6(b)) are displayed. A typical composition of the response upon a linear increasing voltage of a displacement current due to the device capacitance and an additional charge extraction current can be

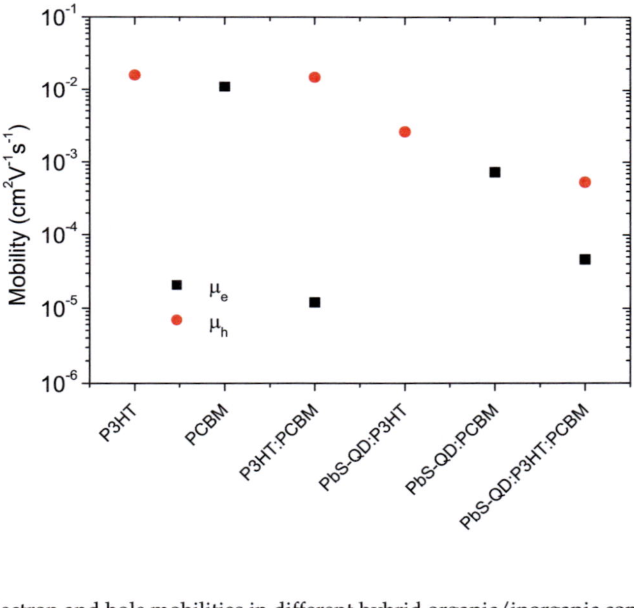

Figure 7.5.: Electron and hole mobilities in different hybrid organic/inorganic composites.

observed. The extraction currents disappear for reverse voltages $V_{Bias} < -5$ V and show maximum values for the measurements at 0.8 V. Under reverse bias conditions the layer might be already depleted or residual charges are too few to create a significant response. Whereas, for voltages from 0 V to 0.8 V the equilibrium charges plus injected charges are present in the layer and contribute to a significant extraction current by a negative voltage ramp of 12.5 V/ms. A slight photodoping during Photo-CELIV measurements can be observed for offset voltages of -0.5 V, 0 V, 0.2 V and 0.4 V - voltages are below diode threshold - as those curves show a pronounced extraction current. Those curves are dominated by photogenerated charge carriers and are lying closer to each other in figure 7.6(b). For this reason, the photogenerated charges contribute distinguishably to the extraction current. A further observation is that with increasing the bias voltage the maximum of the extraction current shifts to the right for both measurements which can be assigned to increased amount of injected charges overlaying the extraction current from equilibrium charges. An indication for that is the strong rise of the extraction current for larger offset voltages as the rise is dependent on the conductivity of the film. Injection of charges from forward bias increases the conductivity as the population of charge carriers in the layer becomes higher.

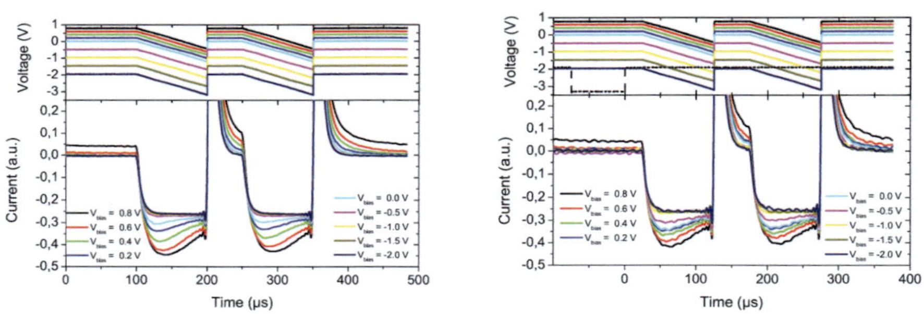

(a) Dark CELIV

(b) NIR Photo-CELIV

Figure 7.6.: CELIV traces at different reverse bias were recorded under dark conditions (**a**) and with NIR illumination at 1310 nm (**b**). Two subsequent voltage sweeps of 100 µs with a ramp of 12.5 V/ms and a time-lag of 50 µs in between. For Photo-CELIV an initial pulse of 75 µs was applied before the voltage sweeps.

The CELIV method was introduced by ref. [71] to analyze charge transport in semiconductors, and it was also applied to investigate bulk mobilities in organic polymer films (cf. ref. [72, 70]). Mobilities can be determined by the following relation:

$$\mu = \frac{2d^2}{3\dot{V}t_{max} \cdot \left(1 + 0.36 \cdot \frac{I_{ext}}{I_{dis}}\right)} \tag{7.1}$$

with d as the thickness, t_{max} as the elapsed time from starting the voltage ramp until the maximum of the extraction current is reached, $\dot{V}$ as the slope of the voltage ramp, I_{ext} as the extraction current and I_{dis} as the displacement current. Figure 7.7 gives the mobilities for dark CELIV (figure 7.7(a)) and Photo-CELIV at 1310 nm (figure 7.7(b)) for the first and second pulse with a delay of 50 µs between the two pulses corresponding to the applied bias and the voltage ramp. No significant and systematic influence for different slopes of the voltage ramp on the extracted mobilities of the equilibrium charge carriers is apparent in the measurements. In figure 7.7(a) values for mobilities between 1.0×10^{-5} cm^2(Vs)$^{-1}$ and 4.3×10^{-5} cm^2(Vs)$^{-1}$ are obtained. Dark CELIV measurements in figure 7.8(a) show even a smaller bandwidth of the mobility from 1.8×10^{-5} cm^2(Vs)$^{-1}$ to 1.6×10^{-5} cm^2(Vs)$^{-1}$ for the different voltage sweeps. Between figure 7.7(a) and 7.7(b) no evident difference can be distinguished, although a slight modification for Photo-CELIV was observed in figure 7.6(b) in comparison to figure 7.6(a) for dark CELIV. For this reason, the amount of photogenerated charges and photogeneration itself do not affect charge transport as similar mobilities are observed for both CELIV measurements. A minor difference of the mobility is found between the first and second CELIV pulse, which can be suspected from figure 7.7 and is clearly visible in figure 7.8. Lower mobilities are obtained for the extraction from the second pulse. Since only slight differences can be observed for the second pulse for the mobilities as well as for the CELIV traces, blocking properties of the top and bottom electrode are pronounced. However, a decrease of the mobility is observed with increasing offset voltage. Due to the injection of charge carriers at forward bias, equation 7.1 is not fully applicable anymore as this equation accounts for equilibrium charge carriers. This can be clearly seen in figure 7.7 and in particular in figure 7.8(b). Note that for large offset voltages the CELIV traces do not show saturation to determine the displacement current exactly, but estimating the displacement current from lower offset voltages introduces only a minor error. Further Photo-CELIV measurements with excitation at 532 nm were performed as well, but no different results were observed.

Ambipolar charge carrier mobilities around a value of 2.0×10^{-5} cm^2(Vs)$^{-1}$ could be determined during CELIV measurements. This is significantly lower as the values obtained by field effect mobility measurements of 6×10^{-4} cm^2(Vs)$^{-1}$ for holes and $\sim 4 \times 10^{-5}$ cm^2(Vs)$^{-1}$ for electrons. Field effect

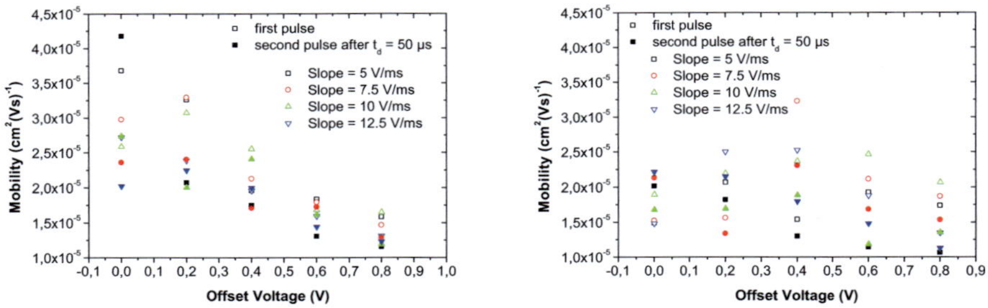

(a) Dark CELIV

(b) NIR Photo-CELIV

Figure 7.7.: Extracted mobilities by dark CELIV (**a**) and Photo-CELIV measurements at 1310 nm (**b**). Neither the slope of the voltage ramp nor the fact that the mobilities are extracted from the first or the second pulse seem to have a major influence. Photo-CELIV measurements also do not show variations beyond tolerance levels.

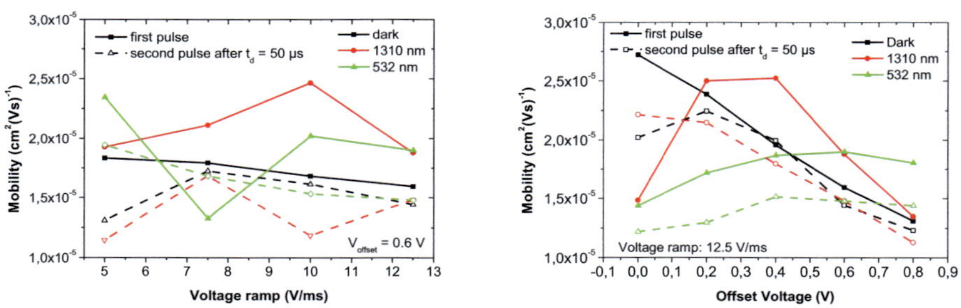

(a) Voltage Ramp

(b) Offset voltage

Figure 7.8.: Extracted CELIV mobilities in dependence on the slope of voltage ramp (**a**) and the applied offset voltage (**b**). Neither for dark, green nor NIR measurements an influence of the slope of the voltage ramp can be observed. A slight decay in mobility with increasing the offset voltage and for the second pulse can be seen in the measurements.

measurements are usually giving overestimated values for the mobility in comparison to bulk mobilities, so that the obtained results seem to be consistent.

Charge transfer analysis was investigated also by pump and probe measurements, but no clear results were found to conclude on how charge transfer occurs in hybrid films.

7.2.2. Luminescence of Hybrid Films

In figure 7.9 the photoluminescence spectra of an IO-HPD are depicted. The radiative decay via the first excitonic state can be observed at 1550 nm (see figure 7.9(a)). The IO-HPD was optically excited by a laser emitting at 935 nm and photoluminescence spectra at different biases were recorded. A minor dependence of the photoluminescence spectra on the applied bias is found (see 7.9(b)). One would assume that with increasing reverse bias the excited charge carriers are withdrawn from the PbS-QDs and the photoluminescence is suppressed which can be observed as small a quenching effect in figure 7.9(b) [61].

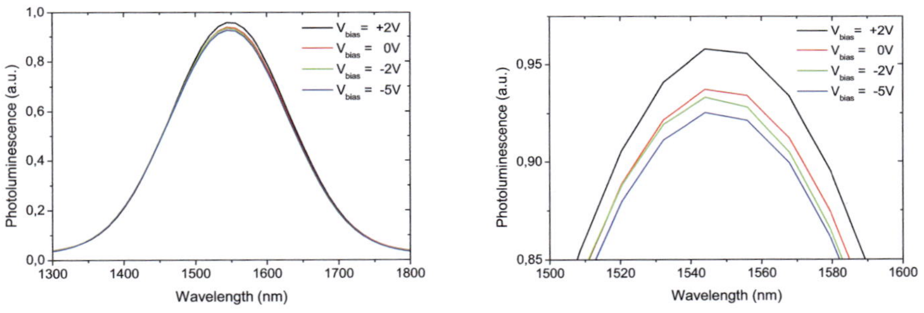

(a) Photoluminescence spectra

(b) Zoom at peak wavelength

Figure 7.9.: Photoluminescence spectra of PbS-QD:P3HT:PCBM devices at different bias voltages (**a**) and zoomed in to the peak wavelengths (**b**).

7.2.3. Spatially Resolved Photocurrents in Hybrid Films

Spatially resolved measurements on IO-HPDs are a possibility to identify variations in photogeneration of charge carriers due to inhomogeneities. In the performed measurements a spot of roughly 3 µm for excitation with a 532 nm laser and 10 µm for a 940 nm NIR LED scans in the active area at steps of 5 µm. Figure 7.10 displays a normalized contour plot for green excitation on the left and for NIR excitation on the right. The green scan shows little variation and more homogeneous behavior than the NIR scan. At green illumination P3HT and PbS-QDs contribute to photogeneration of charge, so that the current is better distributed throughout the active area. PbS-QDs solely absorb at 940 nm and therefore the larger inhomogeneity of the NIR scan can be attributed to morphology (e.g.PbS-QD concentration) and film thickness fluctuations. Due to the low fluctuation of about $\pm 10\,\%$ (see figure 7.11(a)) obtained in the scan, film thickness is assumed to vary little. The variation of up to $\pm 30\,\%$ (see figure 7.11(b)) is therefore probably caused by domains with different morphology. TEM and SEM analysis showed that PbS-QDs are distributed almost uniformly throughout the film. However, some domains show less PbS-QD concentration (cf. figure 7.3(a) and 7.2) and those images represent only a small part of the device. PbS-QD:P3HT:PCBM may vary also for different domains resulting in different charge transfer properties. Togehter with some fluctuations in the film thickness this might lead to the observed variations in the spatial photocurrents. In summary, hybrid photodiodes exhibit a spatial dependent photocurrent due to variations of film properties.

7.3. Electro-optical Characterization of hybrid Photodiodes

IO-HPDs are intended to be used as optoelectronic device for photodetection. For this reason, electro-optical properties of IO-HPD devices are examined, in particular. At first, static behavior and subsequently dynamic characteristics are subject of the investigations.

7.3.1. Static Device Properties of IO-HPDs

As static properties this section deals with temperature dependence, NIR photoresponse and the I-V characteristics. From the results, a possible equivalent circuit diagram is derived.

Temperature Dependence

The temperature dependence of the I-V characteristic can give a picture about the situation of intrinsic densities of traps and defects which contribute gradually to the intrinsic density of free charge carriers and therefore to the diode

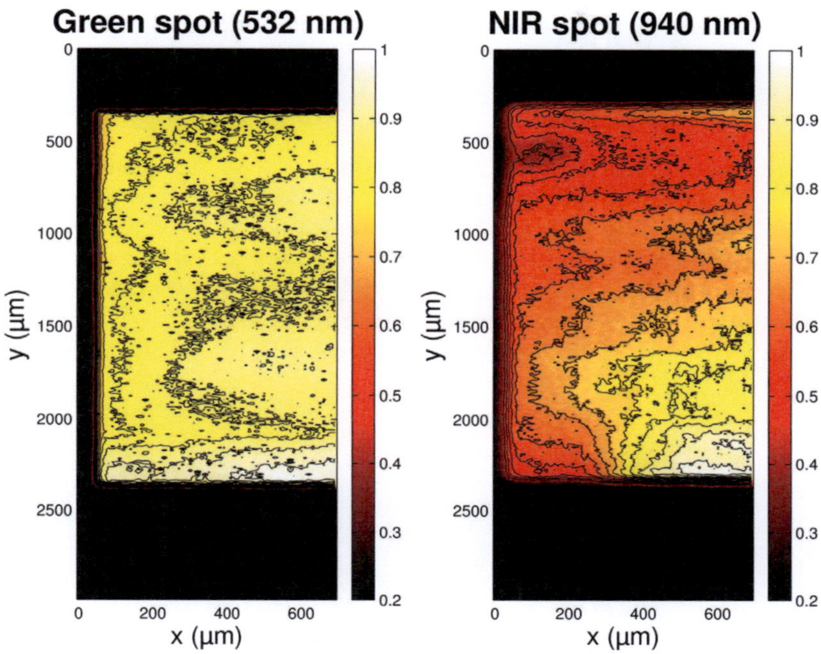

Figure 7.10.: Contour plot of the spatial resolved photocurrents on IO-HPDs with green excitation at 532 nm (**left**) and NIR excitation at 940 nm (**right**).

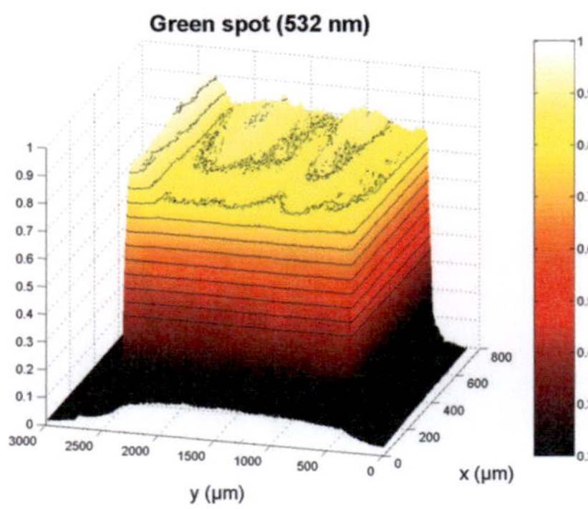

(a) Green scan

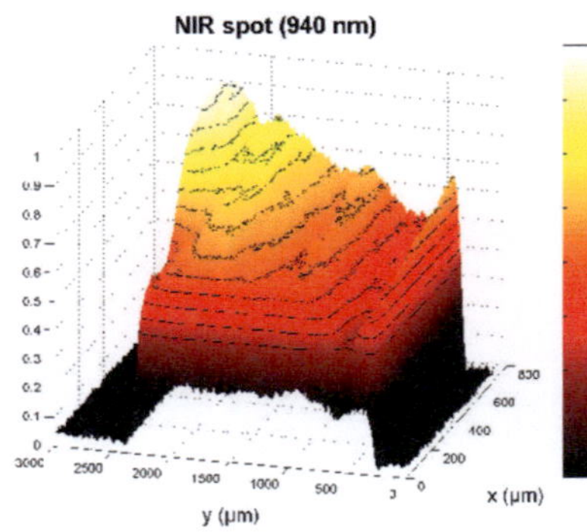

(b) NIR scan

Figure 7.11.: Normalized 3D plot of the spatial resolved photocurrents on IO-HPDs with green excitation at 532 nm (**a**) and NIR excitation at 940 nm (**b**).

current with increasing temperature. Equation 7.2 from semiconductor physics makes apparent that diode currents raise with temperature.

$$I(T) = I_S(T) \cdot \left[exp\left(-\frac{qV}{nk_BT}\right) - 1\right] \tag{7.2}$$

with I as the diode current, I_S the saturation current, q the elementary charge, n the ideality factor, k_B the Boltzmann constant and T as the temperature. Figure 7.12 illustrates the temperature dependence of the I-V characteristics. By increasing the temperature from 0 °C to 80 °C forward and reverse currents grow accordingly by a factor of $\sim$3 and $\sim$25, respectively. Saturation currents can be estimated from the intersection of the extrapolated linear fit with the y-axis at 0 V as indicated by the dashed lines in the figure 7.12. Due to equation

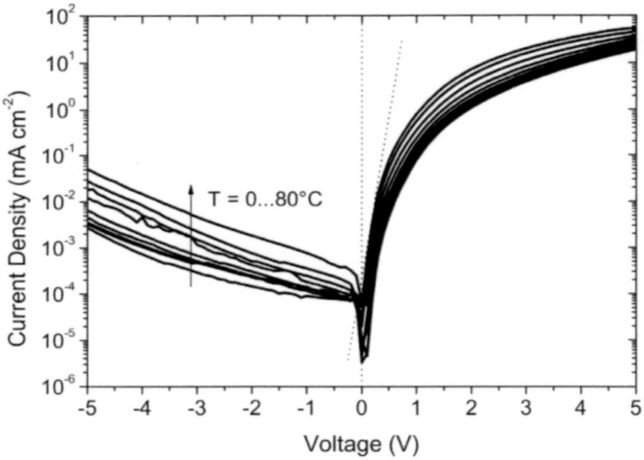

Figure 7.12.: I-V characteristics of an IO-HPD with a temperature sweep from 0 °C to 80 °C. The dashed line indicates to extraction of saturation currents by the intersection of an extrapolated linear fit with the y-axes at 0 V.

7.3 an exponential dependence to $1/T$ is expected also for the saturation current I_0. In inorganic semiconductor diodes saturation currents are also related to the energy bandgap E_g. In a semi-logarithmic plot of saturation currents vs. $1/T$ (Arrhenius plot) the slope reveals the ratio of $\frac{E_g}{nk_B}$. In figure 7.13(a) the relation of the saturation currents obtained by the extrapolated fit for each temperature in figure 7.12 is shown. A slope of 452 meV is found for $\frac{E_g}{n}$ in IO-HPDs.With the ideality factors of 2.31 - 2.44 which can be found in figure 7.12, a value of 452 meV $\cdot$ 2.4 $=$ 1084 meV can be estimated for E_g. The PbS-QDs applied in these devices exhibit a cut-off wavelength for absorption at 1670 nm corresponding to 743 meV. The difference between the HOMO level of P3HT and LUMO of PCBM is 600 meV. The ternary composition allow possibilities for interactions of PbS-QDs, P3HT and PCBM which may affect the effective bandgap of the ternary composition. The ideality factor n already reveals values of $\sim$ 2.4 which are larger than physically feasible. This indicates that the diode current equations might be not fully applicable for IO-HPDs, so that the equation 7.2 can be reduced and for the saturation current I_S the following dependence on the temperature is found according to semiconductor physics

$$I_S(T) = I_{S,T_{max}} \cdot exp\left(-\frac{E_g}{k_B T}\right) \tag{7.3}$$

Leakage currents can occur from unwanted generation of charge carriers from defects or released from traps and injection from the blocking contacts. As found in 7.12, the leakage currents at reverse bias are also strongly dependent on the temperature. Thermal assisted generation of charges from traps and defects as well as injection seems to be a relevant mechanism in hybrid photodiodes. Generation of charge can take place when enough activation energy is provided to lift a charge carrier from trap or defect states to its band level or to inject charge carriers. For this reason, the reverse current $I_{reverse}$ should be proportional to the expression in equation 7.4 taking the activation energy E_A into account. Plotting the reverse current at -4 V in an Arrhenius plot in figure 7.13(b) a slope and thus an activation energy of 326 meV could be extracted.

$$I_{reverse} \propto exp\left(-\frac{E_A}{nk_B T}\right) \tag{7.4}$$

A pronounced temperature dependence of currents does not surprise as films with QDs are known for high trap densities which can contribute to leakage and saturation currents.

Low temperature photoluminescence measurement at the University of Linz showed that the energy level alignment between the involved materials has to be carefully chosen and is essential for an efficient charge separation and transfer in hybrid devices. At low temperatures, the charge transfer is inhibited since the energy difference between the state is not sufficient [127].

NIR Photoresponse

A short section is explicitly devoted to the NIR photoresponse, although results of EQEs were already presented in chapter 6. An important characteristic is the linearity of the photoresponse with relation to the intensity of the illumination,

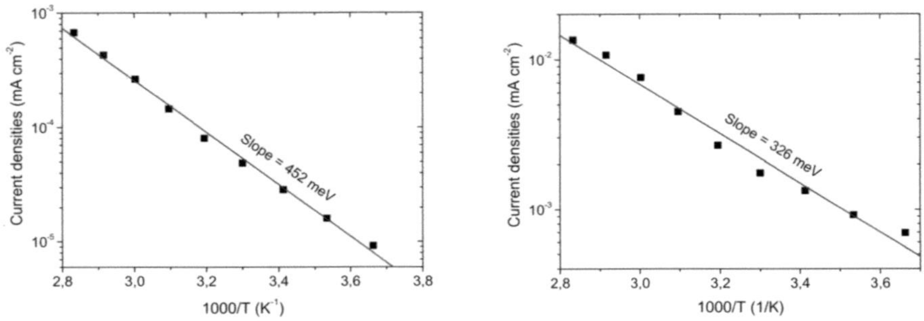

(a) Saturation currents

(b) Dark currents at -4 V

Figure 7.13.: Temperature dependence of saturation and dark currents of IO-HPDs. Reciprocal temperature dependence of (**a**) the saturation current derived from I-V characteristics and (**b**) dark currents at -4 V at temperature from 0 °Cto 80 °C.

which is given in figure 7.14(a). Linearity over more than two decades for intensities from 6 μW cm^{-2} to 800 μW cm^{-2} is achieved by IO-HPDs as the slope of the linear fit in the normalized log-log plot is equal to 1 in figure 7.14(a).

A peculiarity of IO-HPDs is illustrated in figure 7.14(b). EQEs show a strong dependence on the applied bias and make it possible to adjust the sensitivity of IO-HPD. For the investigated device EQEs can be modified linearly in a range from 16% to 51% for voltages from -5 V to -8 V. In the visible at 540 nm a similar behavior is observable. Note that the high EQE values of can be assigned to photoconductive gain of hybrid films. From the linear dependence of the sensitivity on the applied bias and the linear proportionality of the photoconductive gain $G_{pc} = \mu\tau \cdot \frac{V}{d^2}$ a correlation can be assumed.

Determination of Static Device Parameters

In ref. [148, 179] a model to simulate I-V characteristics of OPV devices with good agreement to measurements was introduced. This model was applied with modifications to hybrid photodiodes as dark currents exhibit an additional exponential dependence on the applied bias. This additional current source might be attributed to field assisted transport facilitating trap states in the hybrid film. The modified equation is given by

$$J - J_s \cdot \left[exp\left(q \cdot \frac{V - R_S JA}{nk_B T} \right) - 1 \right] - \frac{V - R_S JA}{AR_P} + J_s \cdot exp\left(\frac{q\,|V|}{\Delta_t} \right) + J_{light} = 0 \tag{7.5}$$

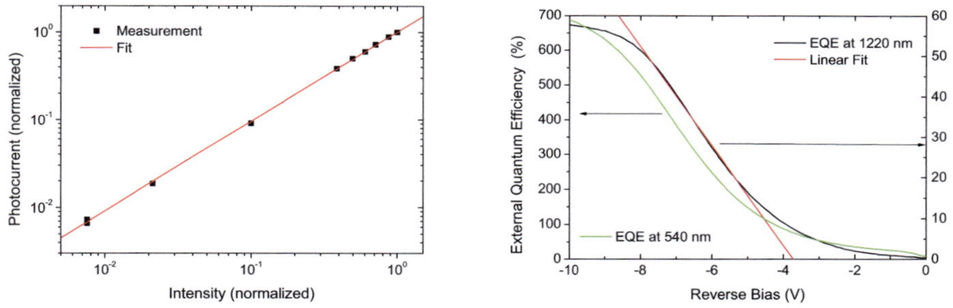

(a) Linearity of Photoresponse

(b) Voltage dependency

Figure 7.14.: Characteristics of NIR photoresponse of IO-HPDs. (**a**) The light current shows a linear dependence on the incident intensity (slope of the linear fit equals 1). Intensities used to measure range from 6 µW cm^{-2} to 800 µW cm^{-2}. (**b**) EQEs at 1220 nm are given with respect to the applied reverse bias. The linear dependence on the bias allows adjustment of the sensitivity of the hybrid device. For comparison, the response at 540 nm is depicted as well.

with the saturation current density J_s , the diode area A, the applied voltage V, the serial resistance R_S, the parallel resistance R_P, the ideality factor n, the Boltzmann constant k_B, the temperature T, the introduced transport energy Δ_t and J_{light} as the photocurrent density. The photocurrent is obtained by

$$j_{light} = \begin{cases} sgn\left(V - V_{bi}\right) \cdot J_{ph} & d > l_{max} \\ \mu\tau \cdot \frac{V - V_{bi}}{d^2} \cdot J_{ph} & d > l_{max} \end{cases} \tag{7.6}$$

with V_{bi} as the built-in potential, J_{ph} the primary photocurrent density, μ the mobility, τ the lifetime and $l_{max} = \mu\tau \cdot (V - V_{bi})/d$ as the schubweg. If the schubweg is larger than the thickness d, all of the primary photocurrent density J_{ph} is extracted. The primary photocurrent is dependent on the absorbed photons creating charge carriers: $\upsilon_{cs} \cdot \frac{P_{opt}}{hf} \cdot (1 - exp\left(\alpha d\right))$. For schubweg smaller than the thickness, only a fraction $\frac{l_{max}}{d}$ of the primary photocurrent can be extracted. During illumination an additional photoshunt is introduced as the parallel resistance is lowered by photodoping ($R_P = R_{P,light} \ll R_{P,dark}$). The photoshunt is expressed by

$$R_{P,light} = 1/\left(\upsilon_{cp} \cdot J_{ph} \cdot \frac{\mu\tau}{d} \cdot A\right) \cdot \frac{A}{d} \tag{7.7}$$

with υ_{cp} accounting for the contact permeation probability.

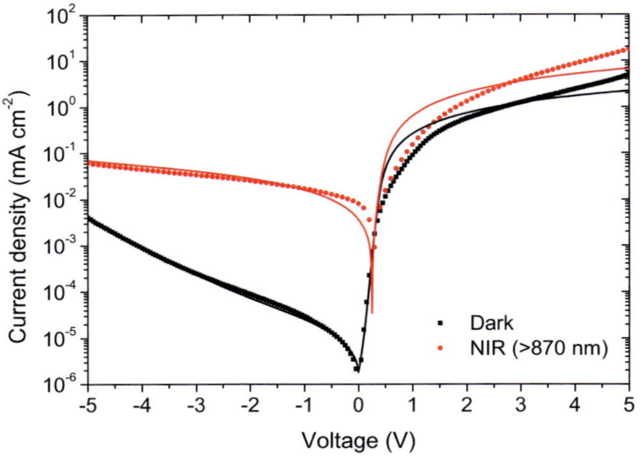

Figure 7.15.: I-V characteristics and the resulting fit of an IO-HPD with PbS-QD absorbing at 1220 nm by the introduced modified model.

According to the introduced model dark and light currents can be reproduced as seen in figure 7.15. Dark currents in reverse bias agree very well with the simulation, but forward dark currents differ significantly from the model as well as forward light currents. Forward currents seem to have a pronounced exponential dependence which is not covered by the model. For light currents, the extraction of photogenerated carriers is also underestimated for reverse voltages $V_{bias} < -1$ V. Since IO-HPDs are usually operated at voltages $V_{bias} < -1$ V, the model is sufficient to describe IO-HPDs under operating conditions for, e.g., design purpose. The best fit of the model was obtain with values for the ideality factor $n = 1.4$, the dark serial resistance $R_{S,dark} = 80$ kΩ, the saturation current $j_0 = 2.0 \times 10^{-6}$ mAcm^{-2}, the mobility $\mu = 1.5 \times 10^{-5}$ cm^2(Vs)$^{-1}$, the lifetime $\tau = 5.0 \times 10^{-6}$ s, the parallel resistivity $\rho_{P,dark} = 3.0 \times 10^{12}$ Ωcm, $v_{cp} = 0.01$ and $V_{bi} = 0.03$.

IO-HPDs can be approximately described by models from OPV devices with minor modifications, but a particular weak spot is the reconstruction of forward currents. However, the good agreement with reverse currents allows to model IO-HPDs for the widely applied operating conditions at reverse bias.

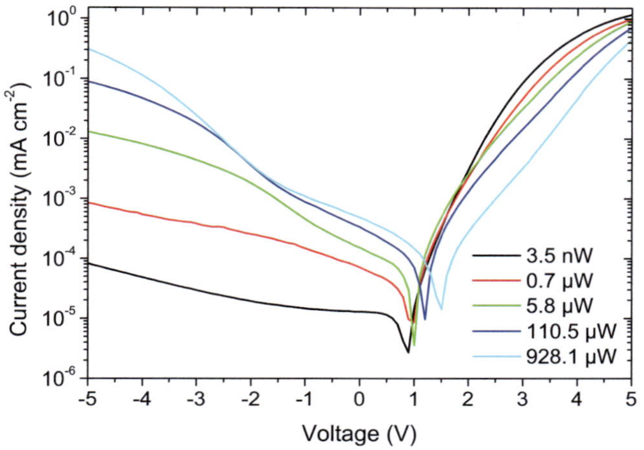

Figure 7.16.: I-V characteristics for the devices used for fast pulsed electro-optical measurements

7.3.2. Dynamic Device Properties of IO-HPDs

The dynamic properties of photodetectors are key figures for their application. Not only high speed communication is limited by the dynamics of detectors, imaging devices need to read-out photodetectors rapidly to acquire images at high frame rates and resolution. In this section IO-HPDs are analyzed by pulsed electro-optical measurements, impedance spectroscopy and noise measurements.

Pulsed Electro-optical Measurements

IO-HPDs were investigated by excitation with short 2.5 ns laser pulse at 532 nm and recording the temporal response of hybrid photodetectors. The I-V characteristics of the analyzed device are depicted in figure 7.16. The response of the IO-HPDs is illustrated in figure 7.17 and 7.18. Figure 7.17 shows the temporal response with relation to the applied bias for intensities of 0.7 µW (figure 7.17(a)) and 928 µW (figure 7.17(b)), while figure 7.18 depicts the response corresponding to the optical density of the excitation at applied bias voltages of -5 V (figure 7.18(a) and -10 V (figure 7.18(b)). Fast rise times of ns and fall times of 60 ns can be observed.

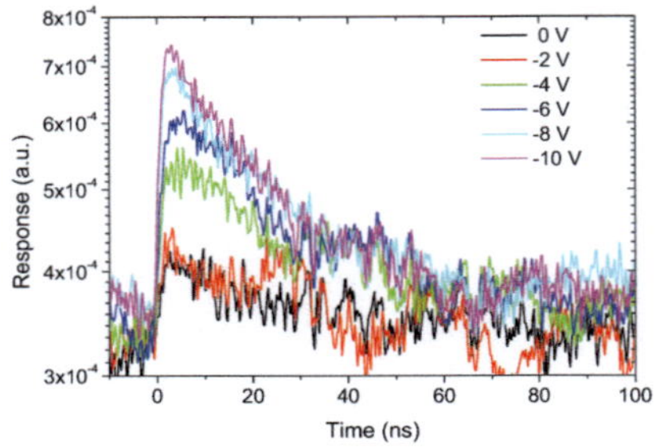

(a) $P_{opt} = 0.7$ µW

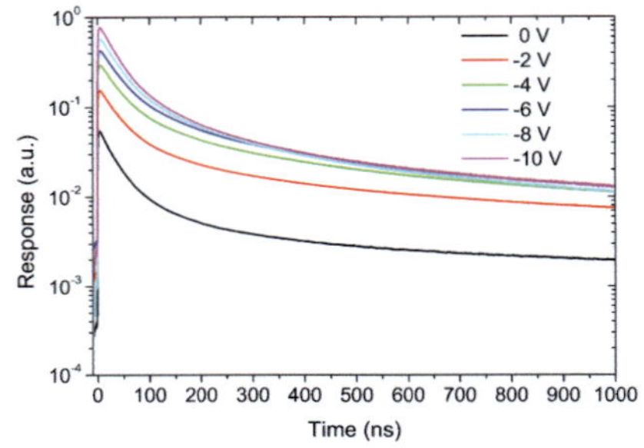

(b) $P_{opt} = 928$ µW

Figure 7.17.: Temporal response for an IO-HPD excited at 532 nm by a laser pulse of 2.5 ns with (**a**) 0.7 µW and (**b**) 928 µW for different voltages.

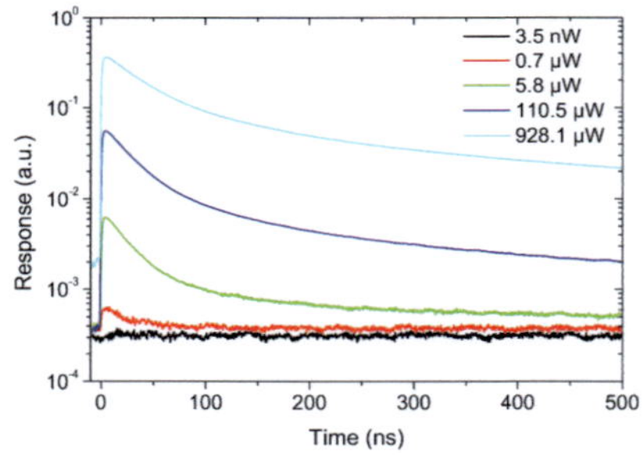

(a) $V_{Bias} = -5$ V

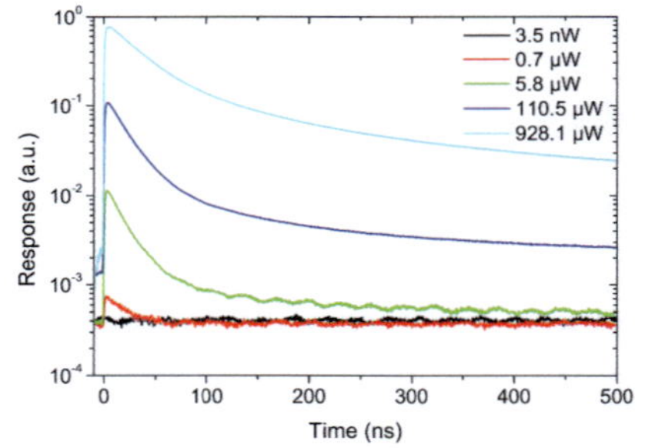

(b) $V_{Bias} = -10$ V

Figure 7.18.: Temporal response for an IO-HPD excited at 532 nm by a laser pulse of 2.5 ns with applied bias voltages of (**a**) -5 V and (**b**) -10 V for optical intensities ranging from 0.7 µW to 928 µW

To investigate closer the decay mechanisms of hybrid devices the temporal response of figure 7.17 and 7.18 were fitted. The decay could not be fitted well by a single exponential decay, but good agreement of the measurements was found with a second order exponential fit. Temporal response functions can be reconstructed well with the following fit equation:

$$I_{ph} = A \cdot exp\left(\frac{t}{\tau_1}\right) + B \cdot exp\left(\frac{t}{\tau_2}\right) + I_{offset} \tag{7.8}$$

with I_{ph} as the photocurrent, A and B as coefficients, t the time, τ_i the respective time constants and I_{offset} as the offset level of the current. The fits for -5 V and -10 V of applied bias at an excitation of 928 µW is given in figure 7.19(a) and 7.19(b), respectively. For fits of the decay at low intensity of 0.7 µW at applied bias voltages of -5 V and -10 V see figure 7.20(a) and 7.20(b), respectively. The parameters of equation 7.8 for each fit are given in table 7.2. As two time constants are found to fit the decay of the response of IO-HPDs well, it is assumed that two decay / recombination processes of photogenerated carriers exist. Both P3HT and PbS-QDs are excited, so that the two time constants probably stem from decay processes related to P3HT and PbS-QDs. Due to different material properties two photoactive materials at 532 nm should exhibit different decays.

(a) $V_{Bias} = -5$ V (b) $V_{Bias} = -10$ V

Figure 7.19.: Fit of the temporal response of IO-HPDs with equation 7.8 for optical intensities of 928 µW at voltages of (**a**) -5 V and (**b**) -10 V. For each fit two time constants were found, but varying with bias voltage. All fitting parameters can be found in table 7.2.

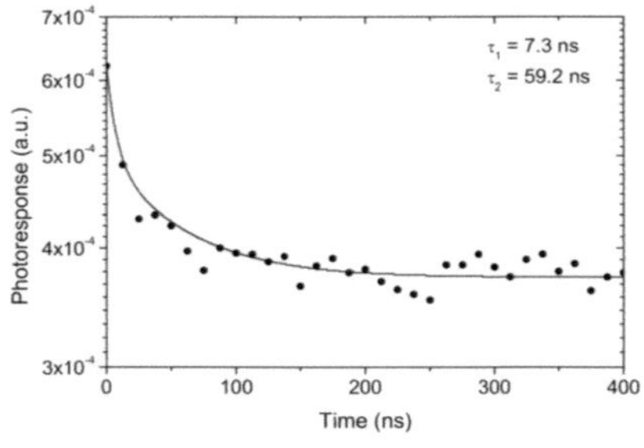

(a) $V_{Bias} = -5$ V

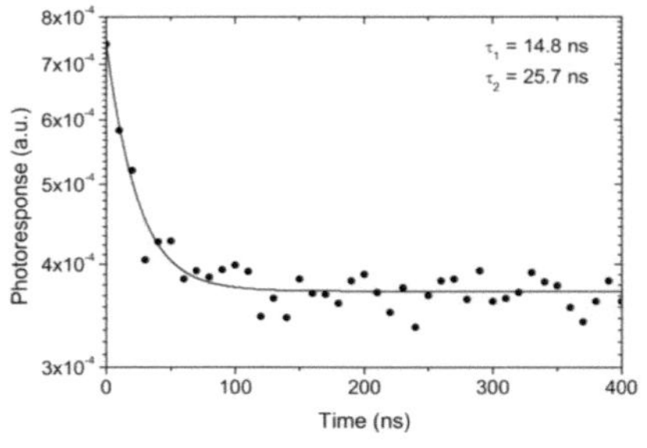

(b) $V_{Bias} = -10$ V

Figure 7.20.: Fit of the temporal response of IO-HPDs with equation 7.8 for optical intensities of 0.7 µW at voltages of (**a**) -5 V and (**b**) -10 V. As observed in figure 7.19 time constants differ with bias voltage. All fitting parameters can be found in table 7.2.

P_{opt}	Bias	A	τ_1	B	τ_2	I_{offset}
0.7 µW	-5 V	1.2×10^{-4}	7.3 ns	1.2×10^{-4}	59.2 ns	3.7×10^{-4}
0.7 µW	-10 V	1.8×10^{-4}	14.8 ns	1.8×10^{-4}	25.7 ns	3.7×10^{-4}
928 µW	-5 V	0.22	34 ns	0.12	148 ns	3.7×10^{-4}
928 µW	-10 V	0.1×10^{-4}	13.9 ns	0.1	30.1 ns	3.7×10^{-4}

Table 7.2.: Parameters used for equation 7.8 to fit the exponential decay of the temporal response of IO-HPDs.

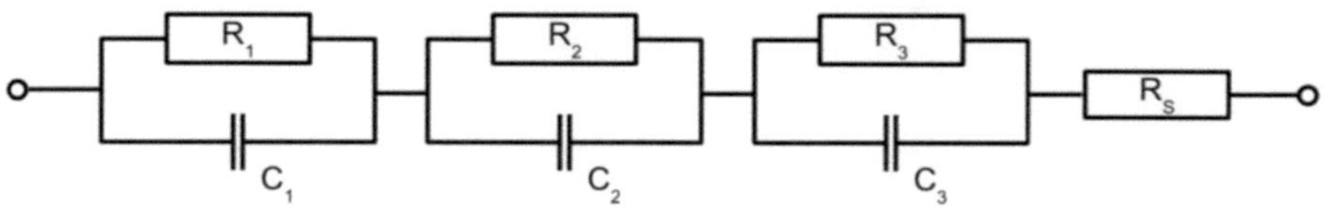

Figure 7.21.: Equivalent circuit used for the description of impedance spectroscopy measurements.

Analysis of Dynamic Components with Impedance Spectroscopy

Impedance spectroscopy measurements were performed to get a better insight to device properties at high frequencies. The impedance spectroscopy measurements makes the global impedance $Z_{tot}(f)$ of a device available which can be approximately described by an equivalent circuit (see figure 7.21). Equations 7.9 to 7.11 express the relation of the total impedance Z_{tot} according to the components of the equivalent circuit in figure 7.21.

$$Z_{tot}(f) = \left(\frac{1}{R_1} + j2\pi f C_1\right)^{-1} + \left(\frac{1}{R_2} + j2\pi f C_2\right)^{-1} + \left(\frac{1}{R_3} + j2\pi f C_3\right)^{-1} + R_S \tag{7.9}$$

$$Re\left\{Z_{tot}(f)\right\} = \frac{R_1}{1 + (2\pi f R_1 C_1)^2} + \frac{R_2}{1 + (2\pi f R_2 C_2)^2} + \frac{R_3}{1 + (2\pi f R_3 C_2)^3} \tag{7.10}$$

$$Im\left\{Z_{tot}(f)\right\} = -2\pi f \left(\frac{C_1 R_1}{1 + (2\pi f R_1^2 C_1)^2} + \frac{C_2 R_2^2}{1 + (2\pi f R_2 C_2)^2} + \frac{C_3 R_3^2}{1 + (2\pi f R_3 C_3)^2}\right) \tag{7.11}$$

Obtained results by impedance spectroscopy for the capacitance ($-1/2 \cdot \pi \cdot f \cdot Im\{Z_{tot}(f)\}$) and resistance ($Re\{Z_{tot}(f)\}$) and the corresponding fits are shown for voltages of 0.0 V and 0.8 V in figure 7.24 and figure 7.23, respectively. These values labeled as $R(f)$ and $C(f)$ converge with increasing applied reverse bias. For OPDs two RC elements were found to fit best (see ref. [187]), whereas for a hybrid photodiode an equivalent circuit with 3 RC elements fit best as depicted in the equivalent circuit of figure 7.21. The fits show a good agreement with the measurement results. Parameters for the components found by the fit are given in table 7.3. Impedance measurements showed values for the resistance at 0.8 V which are consistent with values used for fitting the I-V characteristics (73.4 kΩ + 21.3 kΩ + 3.1 kΩ to 80 kΩ). As already

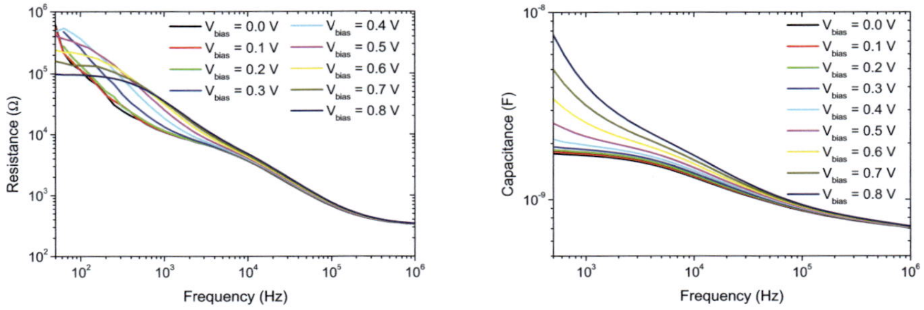

(a) Overall Resistance R(f)

(b) Overall Capacitance C(f)

Figure 7.22.: Obtained overall resistances R(f) (**a**) and capacitances C(f) (**b**) by impedance spectroscopy for voltages from 0.0 V to 0.8 V at steps of 0.1 V.

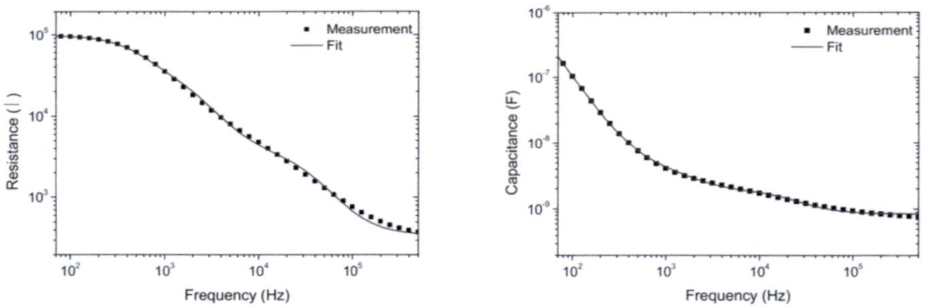

(a) Overall Resistance R(f) at V0.8

(b) Overall Capacitance C(f) at V0.8

Figure 7.23.: Impedance spectroscopy measurements at an applied bias of 0.8 V and the corresponding fit for (**a**) R(f) and (**b**) C(f) by the equivalent circuit depicted in figure 7.21. Fitting parameters for the components are given by table 7.3.

	Bias	R_1	C_1	R_2	C_2	R_3	C_3	R_S
R (f)	0.8 V	73.4 kΩ	4.0 nF	21.3 kΩ	3.0 nF	3.1 kΩ	1.5 nF	0.35 kΩ
C (f)	0.8 V	73.4 kΩ	4.5 nF	21.3 kΩ	3.2 nF	3.1 kΩ	1.5 nF	—
R (f)	0.0 V	2876.9 kΩ	1.8 nF	6.8 kΩ	3.0 nF	0.8 kΩ	2.1 nF	0.35 kΩ
C (f)	0.0 V	2878.8 kΩ	1.8 nF	6.8 kΩ	3.0 nF	0.8 kΩ	2.1 nF	

Table 7.3.: Parameters used for equation 7.8 to fit the exponential decay of the temporal response of IO-HPDs.

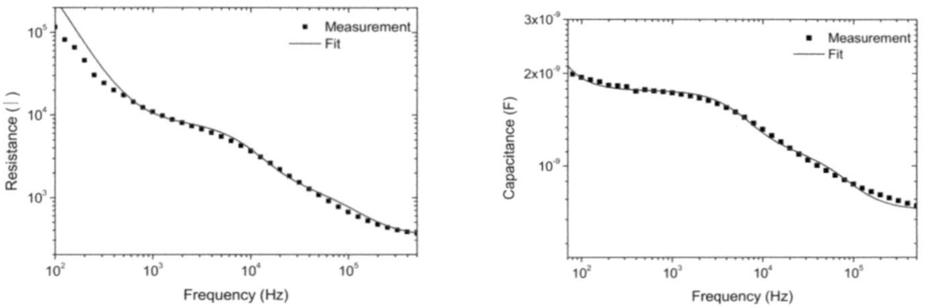

(a) Rf 0V (b) Cf 0V

Figure 7.24.: Impedance spectroscopy measurements at an applied bias of 0.0 V and the corresponding fit for (**a**) R(f) and (**b**) C(f) by the equivalent circuit depicted in figure 7.21. Fitting parameters for the components are given by table 7.3.

observed during electro-optical pulsed measurements PbS-QDs introduce an additional time constant. The influence of the composite and in particular PbS-QD on impedance spectroscopy measurements can be also observed. One can assume that PbS-QDs cause hampered transport conditions which needs to be considered by an additional (third) RC-element in the equivalent circuit.

Noise Analysis of IO-HPDs

Noise properties of detectors have a strong impact on the resolution or the signal which can be detected. In figure 7.25 the noise current over frequency for different reverse bias voltages is given. An increase of the noise current with gradually increasing the reverse bias can be observed (see figure 7.26(a)). Since with increasing reverse biases the dark currents rise consequently as well, one can assume shot noise as major contribution for noise currents. However, the increase of noise currents cannot be explained only by shot noise as depicted in figure 7.26(b). The IO-HPDs are limited by shot noise for low dark currents. The remaining noise current can be assigned to flicker noise. Flicker noise is also dependent on the dark current (c.f. equation 3.8) and is a dominant noise source below frequencies of 10 kHz. For the introduction to noise theory refer to subsection 3.1.2

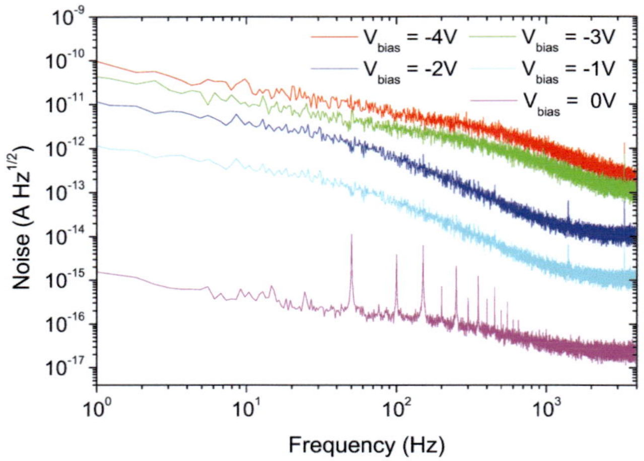

Figure 7.25.: Noise over frequency for different applied reverse voltages.

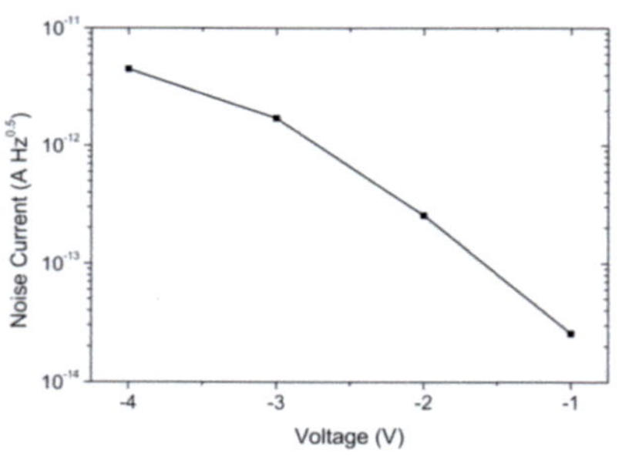

(a) For different applied reverse voltages

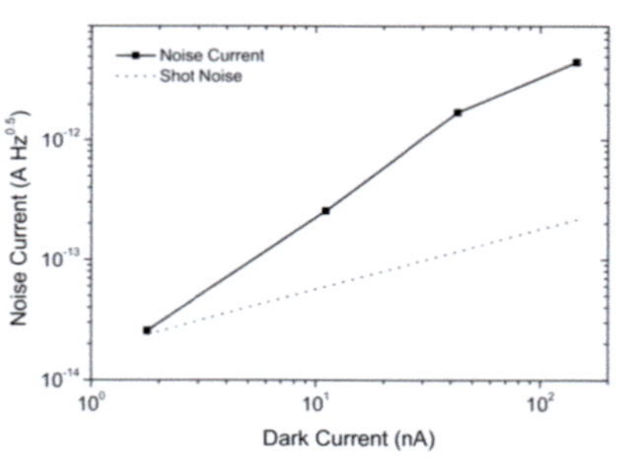

(b) Shot Noise related

Figure 7.26.: Noise current densities at 170 Hz (**a**) for different voltages and (**b**) for different dark current. Shot noise is a limiting factor for the noise.

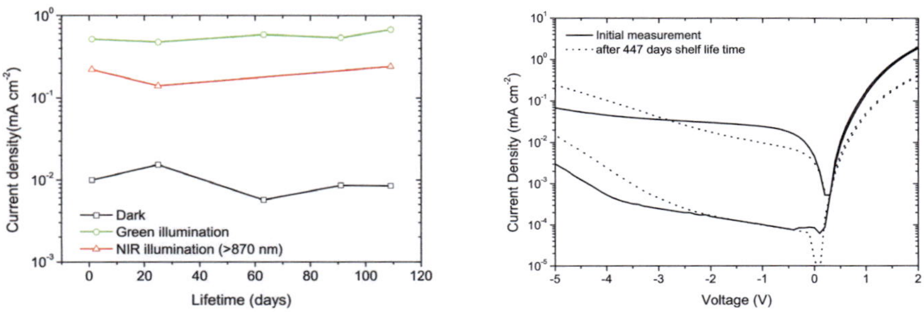

(a) Accelerated Lifetime Measurement (b) Shelf Lifetime Measurement

Figure 7.27.: Measured lifetimes of an IO-HPD (**a**) in a setup for accelerated lifetime measurements and (**b**) for shelf lifetimes.

7.3.3. Stability of IO-HPDs

With respect to applications, lifetime of photodetectors is a very important issue. To prove long lifetime IO-HPDs were investigated in accelerated and shelf lifetime measurements. In figure 7.27(a) an hybrid photodiode was monitored over 110 days regarding dark current, green and NIR light currents. 63 days were conducted under accelerated conditions of 38 °C and 95 rh which corresponds to an acceleration factor of ~6 [124]. The corresponding stability for dark and light currents considering the acceleration factor is more than a year. Shelf lifetime measurements (see figure 7.27(b)) confirmed the results of the accelerated measurements. After 447 days the IO-HPD kept its photodiode characteristics with a moderate increase of dark and light currents. Conductivity properties seem to change over time. Those life-times of more than a year compared to reported values of few weeks [31] might be explained by the oleic acid capping which do not need to be exchanged by shorter ligands for transport reasons. OA ligands insulate and therefore protect PbS-QDs more efficiently than other ligands. An additional protection of PbS-QDs is provided by the surrounding organic polymer matrix which can be seen as well as a protective shell for the PbS-QDs. Due to the oBHJ the PbS-QDs can be embedded as photoactive composite and the organic semiconductor as transport material. For this reasons IO-HPD can benefit from OA ligands and oBHJ by significantly improved lifetimes.

8. Near-Infrared Imaging with PbS-QD sensitized IO-HPDs

Inorganic/organic hybrid photodiodes with embedded PbS-QDs have shown in the recent chapters the potential for near-infrared imaging. With PbS-QDs in oBHJ, the spectral sensitivity was successfully extended up to 1850 nm while preserving the photodiode behavior with dark currents reproducible at 10^{-3} mAcm^{-2} and a bandwidth of 2.5 kHz (-3 dB) was obtained for standard testdiodes (layout see figure 3.2). Note that with an optimized device layout, rise and fall times in the nanosecond-regime can be reached. Furthermore, the photoresponse of IO-HPDs is linearly dependent on the intensity. Thus, IO-HPDs proved to be an alternative to conventional imaging technologies.

Especially for the NIR regime, an increasing number of applications, such as optical tomography [150], process monitoring [6] or night vision [74] are developed and are creating growing interest and demands for new solutions. In particular, the spectral region between 1 µm and 1.8 µm is of commercial interest due to the low water absorption in this range. InGaAs photodiodes are state-of-the-art but expensive solutions for NIR imaging applications and are usually integrated on CMOS based read-out integrated circuits (ROIC) [6]. Alternative approaches to NIR detection include epitaxially grown quantum dots (QDs) and quantum well infrared photodetectors [96, 97, 23]. For the fabrication of large area devices, organic semiconductors can be a suitable alternative to epitaxial structures due to their ease of processing.

8.1. Integration of IO-HPDs on a-Si TFT Imaging Sensors

Since hybrid films with PbS-QDs have similar deposition processes as solution processable organic devices, the fabrication of hybrid imagers is very closely related to organic imagers. As backplane a-Si amorphous silicon (a-Si) panels with an active TFT matrix for pixel addressing can be used. The composite of PbS-QDs:P3HT:PCBM (4:1:1 weight ratio) were coated by doctor-blading ($\sim$ 200 nm) on top after spin-coating of the PEDOT:PSS interlayer ($\sim$ 100 nm).

The 100 nm aluminum top electrode was evaporated subsequently (cf. ref. [25]). The stack of a near-infrared imager with PbS-QDs is shown in figure 8.1(a). The active area of the pixel is defined by the ITO pad, which is designed for a pixel pitch of 154 µm.

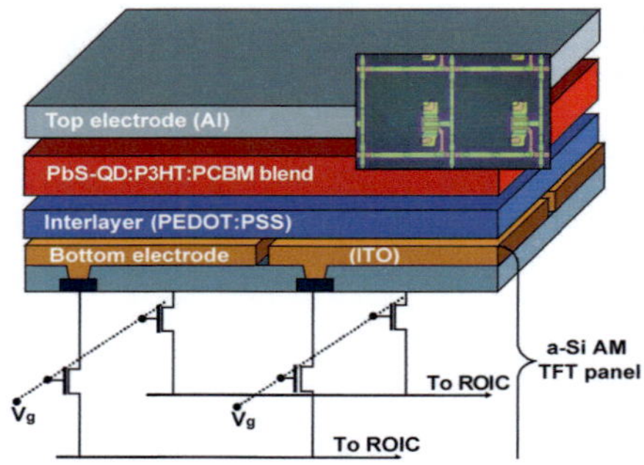

(a) Imager stack

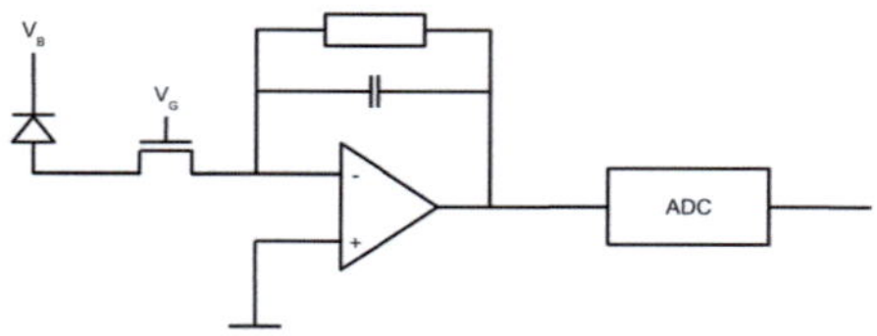

(b) ROIC schematic

Figure 8.1.: Schematic illustration of the imager with an a-Si AM backplane and an unstructured inorganic/organic hybrid frontplane. The inset shows an optical micrograph of two active matrix pixels with a pixel pitch of 154 µm. (**a**) The hybrid photodiode with an active layer of PbS-QD:P3HT:PCBM is processed on top of an a-Si active matrix TFT panel. (**b**) The addressing TFT transfers the signal during readout to the processing circuitry where the collected charges are converted to digital signals by the ROIC.

A simplified cycle of the imaging device to capture an image runs through an exposure part (if the light is not continuous) where the internal pixel capacitance is discharged, followed by the readout with a line time of 20 µs where the corresponding TFTs are closed to transfer the signal to the ROIC. Before starting the acquisition of the next image a reset pulse of 3 ms is applied to bring the pixels to a defined state. During readout the charges from the pixel are collected by a current-voltage operational amplifier and an analog-digital converter. Figure 8.1(b) illustrates a sketch of the ROIC which converts the stored charges in the pixel to a digital signal for further image processing. The ROIC is designed to convert the signal current of a pixel by a 14-bit quantization (16383 counts) corresponding to a quantization level of 4084 electrons for 1 count.

The applied readout circuitry to hybrid photodiodes was originally designed for OPDs in X-ray flat panel detectors, so that experiments were restricted to a bias of max. 5 V and to preconditioned capacitance of the ROIC. However, IO-HPDs showed sufficient performance under these conditions, although the potential of the hybrid imagers could not be fully exploited in terms of quantum efficiency and dark current compensation. To enlarge the collecting time for generated charges in the hybrid photodiode a continuous illumination by a InGaAsP laser diode (ML725B8F) operating at 1310 nm wavelength with a peak intensity of ~ 1.6 mWcm^{-2} was applied. Furthermore, the capacitance of the amplifier was set to 4 pF as a trade-off between sensitivity and dynamic range. High dark currents consumed already a large fraction of the dynamic range of the ADC. Ref. [25] used the same setup for acquiring images by organic imagers for the visible spectrum.

After acquisition a rudimentary image processing was performed. Before capturing the picture, an offset and a bright image was acquired to remove offset structures and to correct inhomogeneities of the illumination and the sensitivity of the pixel by flat fielding. Thus, each pixel of the acquired image is divided by the signal deviation by illumination (bright pixel - dark pixel) after offset correction (acquired pixel - offset pixel). The rudimentary image processing can be summarized as follows in an equation:

$$P_{proc} = \frac{P_{acq} - P_{dark}}{P_{bright} - P_{dark}} \tag{8.1}$$

With P_{dark} as the pixel value without illumination, the offset can be corrected. $P_{bright} - P_{dark}$ gives the rise of the pixel during illumination without any object in the beam. This difference considers the response due to varying pixel sensitivities and inhomogeneities of the light beam for every pixel. $P_{acq} - P_{dark}$ is the offset corrected signal after the acquisition of the image. Equation 8.1 is known as flat fielding. Flat fielding is described by the fact, that if an image is captured without any objects, one should obtain a contrast free image after image processing (see also subsection 3.1.3).

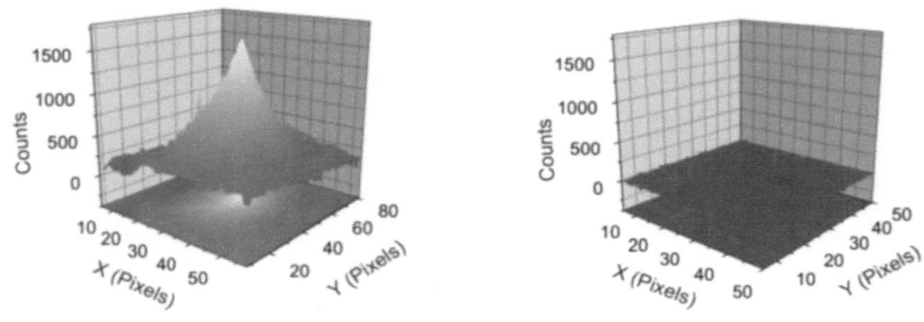

(a) Imager with PbS-QDs:P3HT:PCBM (b) Imager with P3HT:PCBM only

Figure 8.2.: Spot profile images of a focused 1310 nm light source recorded with a P3HT/PCBM reference device (**a**) and with a PbS-QD sensitized IO-HPD (**b**) as absorber layer on the AM backplane. Note, barely any infrared response can be observed with the organic imager without QDs.

8.2. Near-Infrared Imaging

Since all requirements are fulfilled, a near-infrared imaging device with hybrid photodiodes and a processing environment to acquire and to digitize images is available, a deeper look on the characteristics in terms of spatial resolution, crosstalk and further capabilities of the imager can be taken. However, to exclude any parasitic artifacts from a-Si Ge-doping due to near-infrared illumination a comparison of devices with solely P3HT:PCBM and with PbS-QD:P3HT:PCBM in the active layer was undertaken. It is reported that due to trap states a-Si can react sensitive upon NIR illumination. For this reason an illuminated spot from the 1310 nm laser diode was focused on both imagers. In figure 8.2(a) a significant response and a sharp beam profile of the NIR spot on the hybrid imager can be observed. In contrast, the organic imager without PbS-QDs exhibits a flat profile with NIR excitation, only (see figure 8.2). In conclusion, there is no influence on the imager's response from the a-Si panel.

After verifying, that the NIR response is related to the hybrid PbS-QD film, a real image of the solution processed hybrid device by near-infrared illumination can be acquired and analyzed. A slide with a picture showing a monarch butterfly was placed on top of the backplane of the imaging device (ITO electrode faced to the slide) and was illuminated with collimated NIR beam of ~ 35 mm in diameter, so that the object was shadow casted on the imager. The

Figure 8.3.: Near-infrared shadow cast at 1310 nm of a transparent slide showing a monarch butterfly (cf. original slide in the inset).

acquired image after flat fielding and contrast enhancement (windowing) is depicted in figure 8.3. The reconstruction of the picture could be performed already in a high quality, so that further analysis of the images is possible.

To evaluate the crosstalk and resolution a metal sheet with fine structures was attached to the hybrid imager in the optical path of the NIR illumination. The finest details have dimensions of 100 µm, hence smaller than the pixel pitch of 154 µm, but these structures are still significantly visible (see figure 8.4(a)). For measuring crosstalk, the largest rectangle was chosen and the transition from the illuminated domains to the shadowed regions below the metal was analyzed. The drop from 1100 counts to 200 counts in signal level takes place from a pixel to its neighboring pixel. The crosstalk signal disappears after 3 pixels as can be observed in figure 8.4(b). Despite the non-structured hybrid film of the imager, there is only little crosstalk observable.

Another method to analyze crosstalk properties and spatial resolution of the imager is the modulation transfer function (MTF). This function describes the decrease in the signal with the spatial frequency of line patterns. As introduced in chapter 3.1.3 the modulated response of the imager to bar patterns with increasing spatial frequency is monitored. A simplified description of the MTF

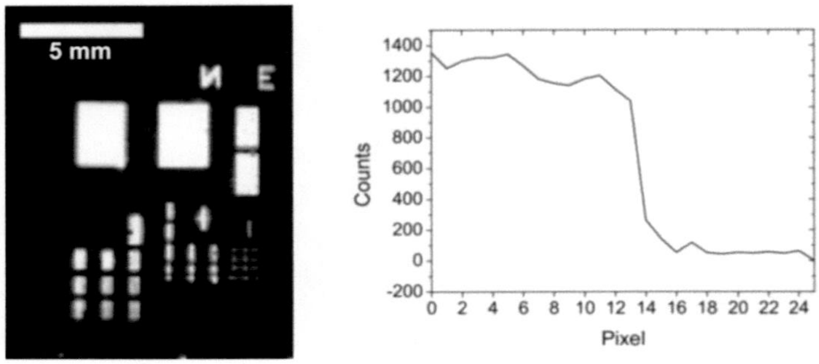

(a) Resolution (b) Crosstalk

Figure 8.4.: A metal sheet with fine structures as small as 100 µm was used for resolution and crosstalk analysis. (**a**) The fine structures are still distinguishable and resolution down to the pixel size is possible. (**b**) The image crosstalk is taken from the response of a pixel row at the sharp metallic edge . The main signal drop occurs within a single pixel and the dark level is reached within 3 pixels.

from ref. [159] was already introduced in chapter 3.1.3 with equation 3.20 and is given by

$$MTF(f) = \frac{A_{bright}(f) - A_{dark}(f)}{A_{bright}(f) + A_{dark}(f)} \tag{8.2}$$

with f as the spatial frequency, A_{bright} as the response in the illuminated part and A_{dark} in the shadowed part of the pattern. For an ideal imager A_{dark} should be always 0 and A_{bright} remains constant, so that the bar patterns can be reconstructed perfectly. But due to imperfections like crosstalk and finitely small pixel pitches, A_{dark} can rise in case of bar patterns coming in the order of the pixel pitch. Pixels below the dark bars can detect signals because of crosstalk and because of being only partially covered by the dark bars. Lead coins with bar patterns of different line pairsmm were attached and the response was analytically analyzed with equation 3.20. The black squares in figure 8.5 is the measured MTF below the lead coins with a linear extrapolation (red line). The shape of the MTF is comparable with a commercial scanner (c.f. ref. [159]) which is not surprising as the images are taken without any optics on the plain panel. The extrapolation intersects with the x-axes at ~ 3.1 line pairs/mm, which is consistent with the pixel pitch of 154 µm, although the decay with higher spatial frequencies is usually not linear.

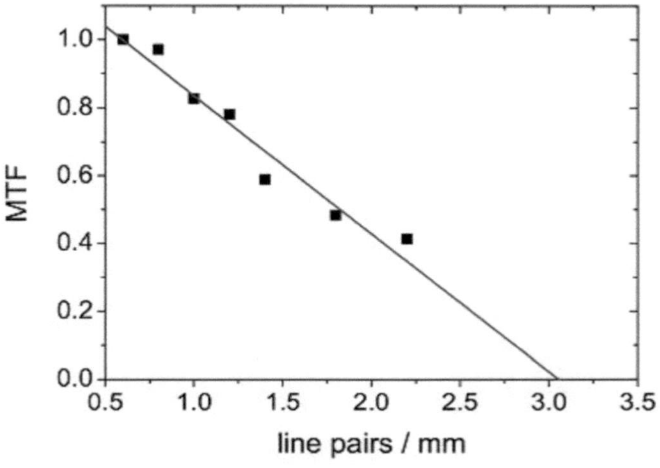

Figure 8.5.: MTF of the imager to define resolution properties. The measurement of lead coins with bar patterns of various spatial frequencies (black squares) with a linear fit (red lines) show a similar behavior comparable with commercial scanners [159].

Finally as the acquisition of single images was well demonstrated, real-time movies were taken with NIR illumination at 1310 nm. Two woodlice (Armadillidium vulgare) were shadow casted on the imager panel recording the movements of the insects. Fine structures during motion, like the cleaning of the antennas of a woodlouse were clearly resolved. Figure 8.6 shows some snapshots from the recorded movie at 5 frames per second. Note that not the imager but the readout electronic limits the frame rate of the imager.

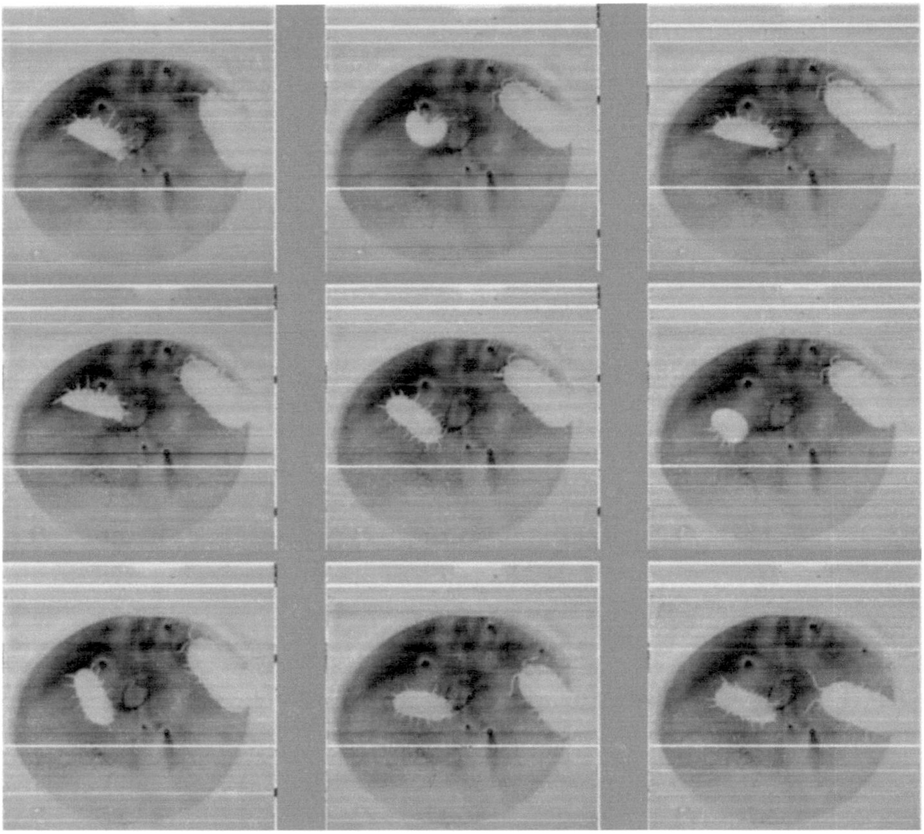

Figure 8.6.: Snapshots from a NIR movie with 5 frames per second at an illumination at 1310 nm showing the shadow cast of two woodlice.

9. X-ray Detection with IO-HPDs

X-ray detection is a key technology for medical imaging, scanning and dosimetry applications. Currently, state-of-the-art technologies use a-Si to detected x-rays by emitted fluorescent light of an excited scintillator. For the ease of flat panel imagers, major effort is spent on developing new technologies for direct conversion of x-rays to replace the conversion from a scintillating film (e.g. amorphous Selenium (a-Se) flat detectors). As the extension of the sensitivity of OPDs to the near-infrared regime by embedding QDs into the polymer matrix was demonstrated, a similar approach might be applicable for X-ray detection. For direct X-ray conversion in the active layer, particles can be chosen accordingly to the two conversion mechanisms of current state to the art technologies. Integrating X-ray fluorescent particles (scintillators) into the polymeric film is called quasi-direct conversion while embedding particles with charge generation from X-rays is designated to direct conversion. Before these two approaches for hybrid organic devices are discussed in detail, a short introduction to important key technologies for X-ray imaging with flat panel detectors is given.

9.1. Conventional X-ray Detection with Flat Panel Detectors

Amorphous silicon flat detectors are already well established on the market for X-ray imaging for over ten years and are currently the first choice of detectors in the latest installed systems [161]. Whereas a-Se flat detectors enter the field at the moment but are still a matter of research. The difference in technology is shown in figure 9.1 indicating the indirect (figure 9.1(a)) and direct conversion (figure 9.1(b)) of X-rays for detection.

9.1.1. Flat Detectors with Indirect X-ray Conversion

As mentioned in the previous section indirect conversion of X-rays is made by an intermediate step of generating visible photons from X-rays with a scintillat-

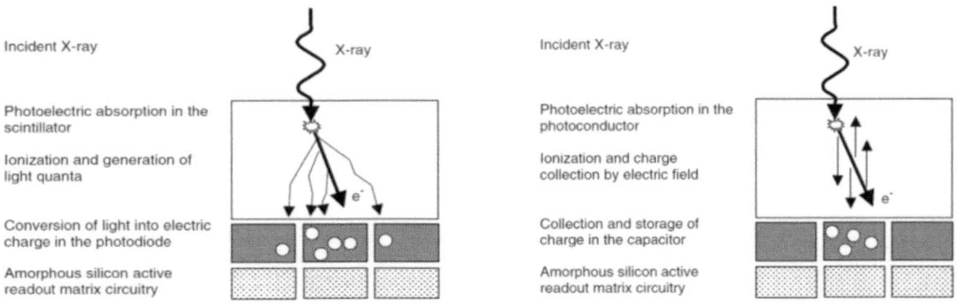

(a) Indirect Conversion (b) Direct Conversion

Figure 9.1.: Indirect and direct conversion of X-rays. (a) A scintillator is generating visible photons due to photoelectric absorption of X-ray quanta. The photons are collected by, e.g., an a-Si photodiode and read out by the active matrix circuitry. (b) Charges are created directly by photoelectric absorption of X-ray quanta which are stored in a capacitor until read out by the active matrix circuitry. Images are reproduced from ref. [161].

ing material which are collected by photodetectors. The photoelectric absorption in the scintillator creates a primary excited electron which dissipates energy to further electrons which relax radiatively by fluorescence. Most widely used scintillators are Cesium Iodide (Cs) or Gadolinium-Oxysulfide (Gd_2O_2S) due to their green fluorescence on X-ray radiation and their processability [161, 107]. In flat panel detectors, the scintillator is grown first and glued on top of an a-Si backplane. The a-Si panel is an active matrix TFT panel with an addressing TFT and a photodiode per pixel where the photons from the scintillation process are detected (see figure 9.2). Typical design parameters of flat detectors are the scintillator thickness and the pixel pitch. With increasing scintillator thickness higher signals can be achieved while the spatial resolution suffers (optical scattering and coupling effects). Analogously, with larger pixel pitches the signal level is increased at the expense of a loss in spatial resolution. Current a-Si X-ray detectors have pixel pitches ranging from 150 μm to 200 μm with scintillator thicknesses varying from 150 μm to 600 μm with respect to the application. Flat detectors with X-ray sensitive areas up to 40 cm × 40 cm are currently commercially available for X-ray imaging [161].

9.1.2. Flat Detectors with Direct X-ray Conversion

Flat detectors utilizing direct conversion of X-rays are highly interesting since the additional step of generating visible light from X-rays can be replaced, thus direct-conversion-imager should be easier and cheaper to manufacture [73, 77].

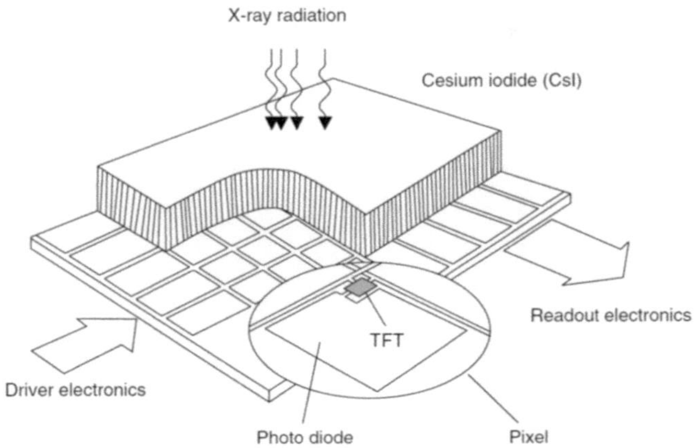

Figure 9.2.: Stack of an a-Si flat detector for X-ray imaging. A scintillator for generation of visible photons from X-ray photons is put on top of an a-Si active matrix TFT panel with a-Si photodiodes. The illustration is reproduced from ref. [161].

As depicted in figure 9.3(a) the photoconductor for direct X-ray conversion is made on top of an active matrix a-Si TFT panel with patterned bottom electrodes defining the pixel size and a non-structured metal top electrode. A closer look to the pixel (see figure 9.3(b)) shows that generated charges by X-rays are driven by an electric field to the electrodes and stored in a capacitor for later read out by the detector circuitry [73]. Material characteristics of suitable direct converters are high X-ray absorption coefficients, low dark currents, efficient extraction of charges and a high generation rate of electrons due to the absorbed X-ray energy. The energy needed for generation of an electron-hole pair (EHP) is therefore a key figure for the X-ray interactions of those materials. The primary generated 'hot' electron dissipates energy while traveling to the electrode creating further EHPs. Furthermore, charge carrier lifetimes and mobilities need to be sufficiently large as the thick layers for X-ray conversion require a long schubweg $\mu\tau F$. The drift velocity μF which is the product of the charge carrier mobility μ and the electric field across the layer d, needs to be large enough to travel during charge carrier lifetime τ to the contacts to extract the generated charge carriers [73, 77, 78]. Lead iodide, lead oxide, cadmium telluride and mercury iodide are applicable materials for direct converters [161, 73]. However, a-Se is mostly used in flat detectors for direct conversion as deposition by thermal evaporation of a-Se on large area substrates is well established. The thickness of a-Se layers depends strongly on the application. While for mammography layer thicknesses of 150 µm to 200 µm are sufficient due to the lower X-ray energy spectrum, for radiographic or fluoroscopic applications thicknesses of 1000 µm are needed [73, 77, 161].

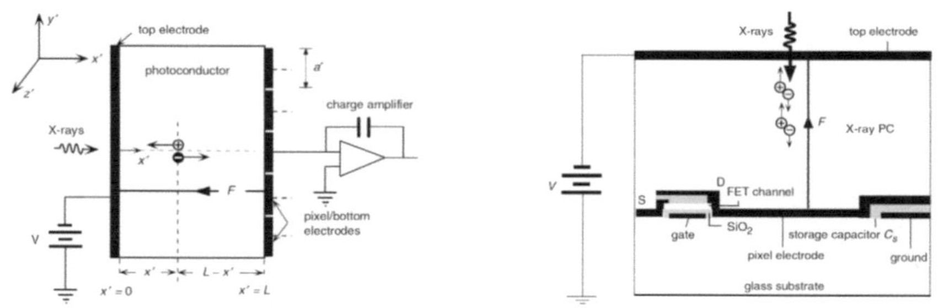

(a) Stack of a-Se flat detectors

(b) Pixel of an a-Se flat detector

Figure 9.3.: Layout of an a-Se flat detector [73]. (a) A-Se is sandwiched between two electrodes. The pixel is defined by the structured bottom electrode. A-Se converts X-rays directly to charges and acts as photoconductor upon X-ray radiation. (b) The generated charges are driven by the applied field to the electrodes and stored in a capacitor on pixel level which can be addressed and read out via an a-Si TFT.

9.1.3. Organic Photodetectors for X-ray Detection

Organic photodetectors have already shown their capability for imaging applications on active matrix a-Si TFT panels [25]. Thus, organic X-ray photodetectors can be designed analogously with similar technologies to a-Si flat detectors with a scintillator on top of the organic photodiodes. Possible benefits for organic photodetectors are that the active layer can be processed from solution with low cost materials over large areas and the active area of the pixel is increased by an additional non-structured organic layer. With a thick layer of up to 30 µm [66] demonstrated a X-ray detection at doses as low as 1.28 mGy/s with an organic semiconductor called poly (riarylamine).

9.2. Embedded X-ray Sensitizers for Direct and Quasi-Direct Conversions in IO-HPDs

In previous chapters (6 to 8) the tailoring of the spectral response by embedding QDs into the active layer was extensively discussed and successfully demonstrated. Therefore applying this approach for the detection of X-ray radiation seems to be promising as well. From current technologies, two alternatives for detecting X-ray radiation might be feasible and are investigated in the following paragraphs: The quasi-direct conversion and the direct conversion of X-rays. Quasi-direct conversions takes into account that the incident X-rays are converted into photons by scintillator particles and absorbed by the oBHJ within a single layer. Whereas direct conversion is due to the fact that nanoparticles (here QDs) absorb X-rays and generate charge carriers from the exposure (see figure 9.4). Note that the following experiments were mainly made to show the proof-of-concept for X-ray detection by hybrid devices.

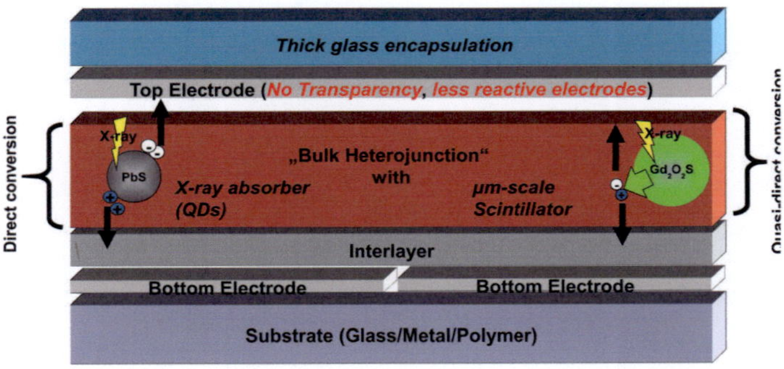

Figure 9.4.: Direct and quasi-direct conversion in IO-HPDs. The QDs convert directly the X-ray, while the scintillator particles transform the X-rays into photons absorbed by the oBHJ.

For both approaches the material requires some additional attributes compared to scintillators or direct conversion materials. Scintillating materials needs to be available in particle size of about several µm in diameter and dispersible in solvents for solution processing. Due to this restrictions and availability the scope of scintillating materials is very restricted. CsI can be excluded as it is grown in 5 µm to 10 µm thick needles [161], so that Gd_2O_2S was selected as it comes as powder and is also frequently chosen as scintillator in state of the art technologies for X-ray imaging. Direct conversion particles shall be soluble and highly X-ray absorbing. Lead compounds have those properties, so that QDs of lead sulfide - the same QDs used for the sensitivity extension of OPDs into the near-infrared - seems to be a good choice, also in terms of availability. In figure 9.5 the mass attenuation coefficients for PbS, Gd_2O_2S, and oleic acid

Material	Density ($g\,cm^{-3}$)	Mass Attenuation Factor @ 30 keV (cm^2/g)	Effective Absorber Mass (g/cm^2)	Penetration Depth (µm)	X-ray conversion	Shell
PbS-QD	7.04	2.52×10^1	3.58×10^{-4}	5.2×10^{-3}	$\propto E_{gap}$	Oleic acid
Gd_2O_2S	7.44	1.16×10^1	1.56×10^{-4}	1.3×10^{-3}	60 photons/keV	n.a.

Table 9.1.: Relevant properties of applied materials for quasi-direct (Gd_2O_2S) and direct conversion (PbS) of X-rays in hybrid devices. Density and mass attenuation factors of the materials were taken from ref. [37].

(ligand of PbS-QDs) is presented for the photoelectric absorption with respect to a 50 kV spectrum. The mass attenuation coefficient α/ρ is the absorption coefficient in relation to the density of the material. Values for the mass attenuation coefficient of the materials are taken from the database of the National Institute for Standards and Technology [37]. PbS shows a slightly higher photoelectric absorption in comparison to Gd_2O_2S. The absorption of oleic acid in the relevant energy range is negligible (<3 orders of magnitude). Furthermore an overview of the selected properties of those materials is given in table 9.1.

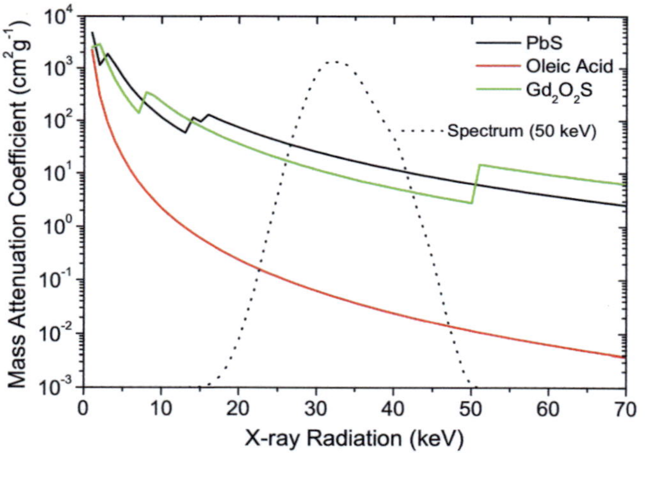

Figure 9.5.: The mass-attenuation factor for the photoelectric absorption of Gd_2O_2S and PbS is shown with respect to a 50 kV spectrum. The neglect of the photoelectric absorption of oleic acid can be clearly seen. Both materials have inherently X-ray absorption coefficients in the same order of magnitude.

To investigate X-ray sensitivity of quasi-direct conversion and direct conversion devices, hybrid films were deposited from solution by spray-coating between two electrodes. On those devices a chopped X-ray beam at a fixed frequency was applied and the modulated response was measured with lock-in technique. For more details of the measurement see section 3.1.1.

$$I_{Signal} = I_{ph}\left(\frac{\tau}{t_r}\right) \overset{t_r=d/\mu F}{=} I_{ph}\left(\frac{\mu\tau F}{d}\right) \tag{9.1}$$

$$I_{ph} = \gamma_{conv} \cdot \int_0^{E_{peak}} \underbrace{\left(1 - exp\left(-\frac{\alpha(E)}{\rho}\cdot\frac{m}{A}\right)\right)}_{X-ray\ absorption} \cdot S_{X-ray}(E)\delta E \tag{9.2}$$

For reconstructing the photoresponse, a simple model considering the relation between the charge carrier lifetime τ and the transit time t_r or between the schubweg $\mu\tau F$ and the layer thickness d, respectively was taken into account. This relation is well known from organic photovoltaics for recombination as long as the schubweg is shorter than the thickness of the active layer [148, 179, 180] and from conventional photoconductors with photoconductive gain [171]. The model is given in equation 9.1 with I_{ph} as the primary photocurrent induced by the X-ray radiation, μ the charge carrier mobility, τ the charge carrier lifetime, F the electric field across the active layer and d the thickness of the active layer. Since an in-depth-analysis of mobility and lifetime have not been performed for each carrier type, lifetime and mobility are not distinguished for holes and electrons. By this model both, recombination and photoconductive gain can be applied with the same relation of the schubweg (lifetime) and thickness (transit time). The model is considered to be applicable as the thick devices are dominated by photoconductivity for the extraction of the devices.

The primary photocurrent in equation 9.2 is described by the photoelectrically absorbed energy of the X-rays with a conversion factor γ_{conv} for generating charge carriers (direct conversion) or photons (quasi-direct conversion), respectively. For e.g. Gd_2O_2S a conversion factor γ_{conv} of 60 photons/keV is reported in literature [107], while for PbS γ_{conv} is determined by fitting. The absorbed energy is defined by the overlap integral of the absorption (dependent on the mass attenuation factor $\alpha(E)/\rho$, the absorbing mass m over the area A and the energy spectrum of the X-rays $S_{X-ray}(E)$.

9.2.1. Quasi Direct Conversion with Embedded Scintillator Particles

The first approach that is investigated is the quasi-direct conversion of X-rays by embedding scintillator particles in oBHJ films. Devices with thick active layers (roughly 10 µm) have been fabricated to ensure sufficient photoelectric absorption of X-rays while remaining thin enough for charge extraction. As depicted in figure 9.4 the active layer with scintillating particles of Gd_2O_2S is sandwiched between a bottom electrode (ITO), an interlayer (PEDOT:PSS) and a top electrode (Al). The deposition was not performed by simple spray-coating, but by co-spray-coating. From two air-guns, Gd_2O_2S-particles dispersed in isopropyl and the oBHJ of P3HT and PCBM dissolved in xylene were spray-coated simultaneously on the substrates. The fraction of Gd_2O_2S-particles of the device is $\sim$ 50 vol % and the active area $\sim$ 50 mm^2. Note that the devices were sealed hermetically under nitrogen atmosphere after processing.

The measurement setup for the X-ray response of hybrid devices is quite similar as known from modulated optical measurements using lock-in technique, despite the source and the modulator. The source was an X-ray generator operating at 50 kV acceleration voltage and the modulator was a rotating chopper wheel with an additional 0.25 mm layer of lead for X-ray absorption. Note that the absorption of this lead sheet was 98.3% for a 50 kV spectrum. Therefore a sufficient attenuation for modulating the X-ray beam was achieved.

(a) I-V characterisitcs

(b) Stability at 100 V

Figure 9.6.: Characteristics of a hybrid device with scintillator particles. (**a**) I-V characteristics of 8 devices with 9.6 µm hybrid film of P3HT:PCBM with $\sim$ 50 vol % of Gd_2O_2S-particles in dark conditions (black lines) and under illumination with ultraviolet light of 365 nm (red lines).(**b**) Stability of dark (black dots) and light currents (red dots) for 100 subsequent measurements at 100 V.

By implementing Gd_2O_2S-particles in thick oBHJ films a photoconductive device without diode properties is obtained (see figure 9.6). For this reason, the model underlying equations 9.1 and 9.2 is applicable. With illumination by an UV source the photoconductive behavior can be clearly observed in figure 9.6(a). Constant and stable dark and light currents (see figure 9.6(b)) are observed.

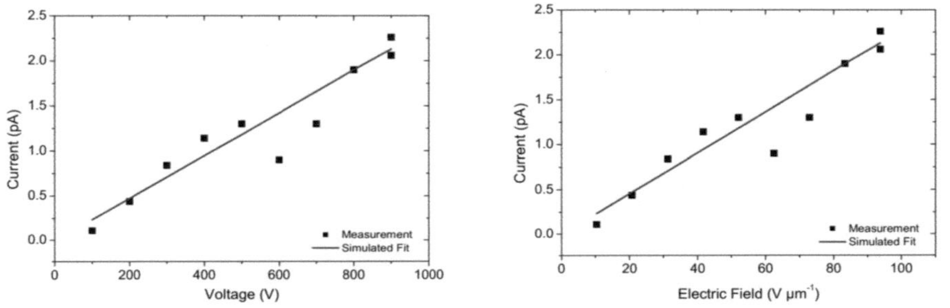

(a) Response vs. applied voltage (b) Response vs. applied electric field

Figure 9.7.: X-ray response of hybrid photodiodes with embedded Gd_2O_2S particles. The 4 mm^2 devices were exposed to a 50 kV X-ray spectrum of ~200 µGy/s. Milling of Gd_2O_2S particles reduced the conversion of X-rays to 6 photons/keV. The modulated response (black squares) of the current is plotted versus the applied bias (a) and the applied electric field (b). The red line is the fitted curve from equations 9.1 and 9.2 with a $\mu\tau$ product of 8.5×10^{-11} cm^2V^{-1}.

Figure 9.7 shows the response of hybrid photodiodes sensitized by embedded Gd_2O_2S particles with an active area of 4 mm^2 to an incident X-ray flux with a 50 kV spectrum (see figure 9.5and a dose of $\sim$ 200 µGy/s. To reduce the particle size for a smoother integration into the organic film, Gd_2O_2S particles were milled for 2 days to reduce the average diameter of the Gd_2O_2S particles. However, continuous milling leads to crystal defects and thus lower scintillation efficiency of Gd_2O_2S. For this reason a loss factor was considered for the conversion and for the experiment 6 photons/keV for X-ray conversion was taken into account [1]. Furthermore, it is assumed, that all photons were absorbed in the 9.6 µm oBHJ layer. Gd_2O_2S particles were very likely surrounded by more than 200 nm of oBHJ. With increasing bias (see figure 9.7(a)) or electric field (see figure 9.7(b)) the response to X-ray can be increased as more and more charge carriers are extracted before recombination (enhancement of the schubweg). At a bias of 900 V (equally to 94 $V\mu m^{-1}$) 2.1 pA of a theoretical estimated maximum photocurrent of 12.8 pA can be measured at the lock-in

[1] private communication with Hr. Fuchs, Siemens Healthcare

amplifier. Note that the theoretical value was estimated from the calculated X-ray absorption of the volume content of Gd_2O_2S particles in the film (50 vol %). The characteristics of the 9.6 µm device is shown in table 9.2 together with the extrapolation for 1000 µm devices. The low signals can be designated to the low absorbed X-ray energy and the reduced scintillation.

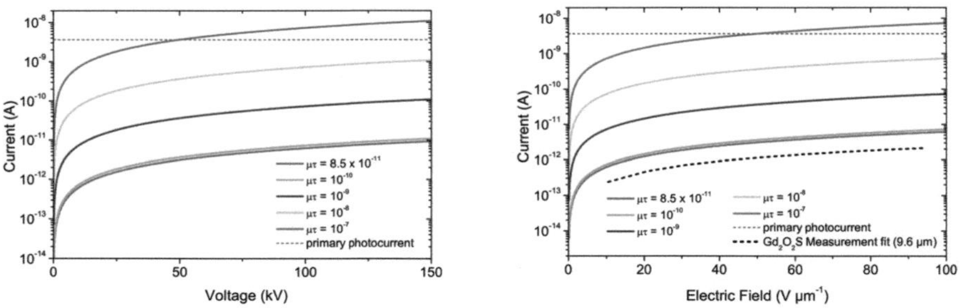

(a) Voltage applied to the active layer

(b) Electric Field across the active layer

Figure 9.8.: Interpolating of 4 mm^2 Gd_2O_2S devices for a layer thickness of 1000 µm with the fitting equations of 9.1 and 9.2. While the parameters for μ and τ extracted from the measurements of the 9.6 µm device show only a little increase for the extracted current, although a loss factor for the scintillation was not considered. Thus, an optimization of the layer morphology to increase lifetime and mobility of the charge carriers is strongly needed as with a sufficient $\mu\tau$ product and with a consequently longer schubweg the generated carriers can be extracted.

In any case, X-ray detection usually deals with low signals because of restrictions on applied doses. To evaluate the potential of quasi direct-conversion with hybrid photodiodes, a interpolation was made for devices with films of 1000 µm with the equations from 9.1 and 9.2. Identical conditions as for the measurements of the 9.6 µm thick devices were considered, but no loss for the scintillation was taken into account. In thick films milling of the scintillator should be avoided with regards to conversion efficiency. Thus, the conversion factor γ_{conv} has the initial value of 60 photons/keV. The results of the simulations are plotted in figure 9.8. Although the thick films rise the absorption of the radiation energy from 3.3×10^6 keV to 9.4×10^7 keV in a photodetector, this gain is lost due to the recombination of the charge carriers before extraction. If the same electric field of 94 Vμm^{-1} (cf. 9.6 µm sample) is applied to the device, 5.8 pA can be extracted. Note that with a reduced scintillation efficiency 2.1 pA were measured experimentally for the 9.6 µm device. Those thick films require long schubwegs, which can be achieved by increasing the electric field or by optimization of material parameters and film morphologies. Since further increasing the electric field is not really feasible, characteristics in terms of lifetime

Layer thickness	Photoel. absorption	Absorption. energy	Photon output	μ (cm^2(Vs)$^{-1}$)	τ (µs)	Primary photocurr.	Response	Elec. field
9.6 µm	3.6 %	3.3×10^6 keV	2.0×10^7	1.6×10^{-5}	5	12.8 pA	2.3 pA	94 Vµm^{-1}
1000 µm	92.5 %	9.4×10^7 keV	5.6×10^9	1.6×10^{-5}	5	3.6 nA	5.5 pA	94 Vµm^{-1}
1000 µm	92.5 %	9.4×10^7 keV	5.6×10^9	10^{-3}	100	3.6 nA	3.6 nA	45 Vµm^{-1}

Table 9.2.: X-ray response for the different layer thicknesses of Gd_2O_2S devices are given with the corresponding parameters. The device with 9.6 µm layer thickness are values obtained from the fit, whereas for the 1000 µm layer the values were projected with the model of 9.1 and 9.2 for the current and for the optimistic situation with parameters from oBHJ films. Note that due to inappropriate handling during milling of Gd_2O_2S particles a loss factor of the scintillation efficiency of 0.1 was taken into account resulting in a conversion factor of 6 photons/keV for the 9.6 µm device.

and mobility need to be dramatically increased to achieve competitive hybrid photodiodes. In table 9.2 the extracted signal currents makes the necessity for optimization obvious. Films with comparable lifetime and mobilities ($\mu\tau$ product of 10^{-7}) to current oBHJ films [42] a full extraction of the signal current 3.6 nA can be achieved at an voltage of 45 kV corresponding to an electric field of 45 Vµm^{-1}. To compare, a-Se direct converting detectors work at electric fields of 10 Vµm^{-1}[78]. In conclusion with further optimizing layers hybrid photodiodes seem to have the potential for alternative direct converters.

9.2.2. Direct Conversion with Embedded Lead Sulfide Quantum Dots

For direct conversion of X-rays in hybrid photodiodes colloidal QDs of PbS are attractive due to the good absorption properties of lead and the demonstrated processability of PbS-QDs in combination with organic films. Hybrid photodiodes of PbS-QDs in a bulk-heterojunction of P3HT:PCBM have shown good results for near-infrared imaging, so that this material combination with the opportunity of industrially scalable production was an evident choice. To detect efficiently X-rays, a multiplication of the absorbed X-ray quanta is necessary by the conversion layer in terms of generated charge carriers. From PbS-QDs and PbSe-QDs, a multiple exciton generation by photons with the energy of several bandgaps was reported by several groups [144, 81, 128, 163]. Though the exact nature of multi exciton generation is still in debate and currently no conclusive picture of this phenomenon has been reported. A strong dependency on the bandgap of the QDs and exciton multiplication factor was found [144, 163]. Therefore, PbS-QDs have the ability to convert excess energy of photons into charge carriers or generate charge carriers with respect to the photon's energy, respectively. From experiments in literature it is assumed that the multiplication process is faster than competing processes like auger recombination, phonon assisted cooling, trapping, etc, and can contribute to charge generation [39, 163]. Nevertheless, ref. [39] describes that impact ionization in QDs can play a role, especially with higher energy photons (c.f. hot electrons).

In contrast to near-infrared detection much thicker active layers to absorb sufficient X-ray radiation have to be fabricated. To reach those thick layers, the spray-coating technique of PbS-QDs:P3HT:PCBM solved in chlorobenzene was applied several times consecutively. Note that the devices were fabricated in a similar way as the samples with Gd_2O_2S particles The weight ratio of the PbS-QDs:P3HT:PCBM film is kept constant to the NIR sensitive films of 4:1:1. In figure 9.9(a) a TEM image illustrates the morphology of thick spray-coated films of PbS-QD:P3HT:PCBM. The PbS-QDs are embedded throughout the whole film in a distinct pattern. These lamellar structure as a result of coalescent drops from the spray coating process have sizes of ~ 100 nm. This quasi homogeneous distribution shall also provide an homogeneous charge generation in the hybrid film, so that the photoconductivity of the film can be increased with X-ray exposure. For layers in the µm-range, the diode characteristic is still conserved which illustrates the figure 9.9(b) with the I-V characteristics under dark conditions and illumination. For X-ray measurements, those devices were investigated in the same setup as introduced for the quasi-direct conversion devices.

To draw conclusions on the efficiency of direct conversion hybrid photodiodes, the X-ray absorption of PbS-QD:P3HT:PCBM films need to be estimated. Since the PbS-core of the QD is the main component for X-ray absorption in a 50 kV spectrum, the fraction of PbS in the active layer is calculated. PbS-QDs are considered to be spherical with a diameter of 5.2 nm which is related to the first excitonic peak in the absorption spectrum [15] and with 14 Å long oleic acid ligands [94]. The binding surface of those ligands of 4.4 Å was estimated by ref. [94]. With the density of PbS (7.6 g/cm^3) and the molar mass of oleic acid (282.46 g/mol), so that the applied PbS-QDs have a density of 2.85 g/cm^3. In a standard hybrid layer with 66.7 wt% of PbS-QDs, 16.7 wt% of P3HT and 16.7 wt% of PCBM, 49.7 vol% are PbS-QD with 1.3 g/cm^3 for P3HT [36] and 1.5 g/cm^3 for PCBM [44]. As the active layer of the device was 2.7 µm thick, an area density of 2.80 g/cm^2 for PbS in the layer was available for X-ray absorption. The organic components were neglected as the absorption is two magnitudes lower than for PbS (cf. figure 9.5). For the primary photocurrent I_{ph}, the photoelectric absorption of PbS in terms of absorbed energy is considered in the model in equation 9.2 by the effective content of PbS-QDs in the film. The conversion factor γ_{conv} of the hybrid layer for X-rays is given in electrons/keV and fitted from real measurements in correlation to the work on multi-exciton generation. Note that this model makes the assumption that the layer's conduc-

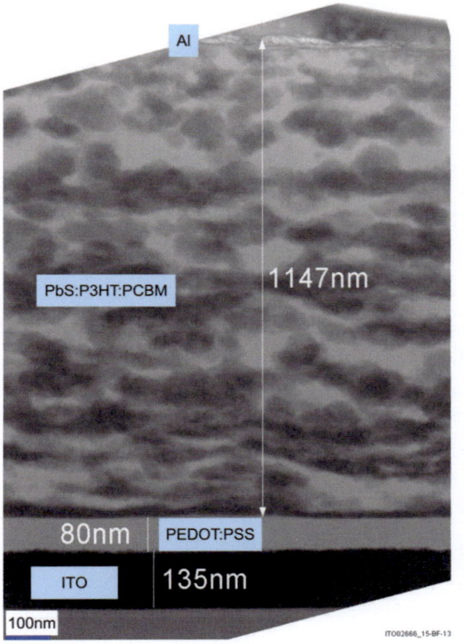

(a) TEM cross section image

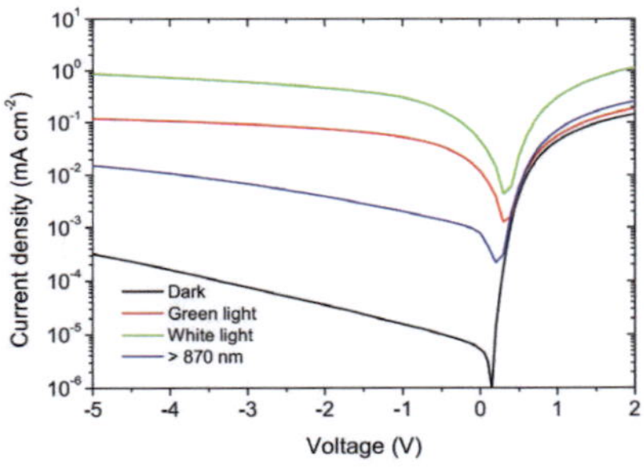

(b) Voltage applied to the active layer

Figure 9.9.: (**a**) TEM image of a spray-coated PbS:P3HT:PCBM hybrid photodiode. The active layer is sandwiched between the bottom electrode (ITO) with the interlayer (PEDOT:PSS) and the top electrode (Al). Three layers have been sprayed one after another to achieve a thickness of 1147 nm. The PbS-QDs are distributed throughout the hole film giving the impression of a lamellar structure. (**b**) I-V characteristics of a 2.7 µm device under dark conditions and for several different light sources.

tivity is changed by X-ray exposure according to photoconductor conditions and relying on the multi-exciton generation in PbS-QDs. Material parameters for PbS and the oleic acid were taken from ref. [95].

In figure 9.10 the response and the fit from a hybrid device with embedded PbS-QDs with an active layer of 2.7 µm to an X-ray beam of ~ 200 µGy/s at 50 kV is depicted with respect to the applied voltage (figure 9.10(a)) and the electric field (figure 9.10(b)). Fitting parameters were 250 electrons/keV for the conversion factor $\gamma_{conv,PbS}$, μ was 1.1×10^{-5} and $\tau = 1.37$ µs ($\mu\tau = 1.5 \times 10^{-11}$ cm^2/V). The conversion factor γ_{conv} was roughly estimated with respect to the PbS-QD's bandgap. To create an electron hole pair by X-ray quanta $2.2E_g + E_{phonon}$ are required at least, which in reality gives a range of $\sim 2.5 \ldots \sim 4E_g$ for the required energy in bulky materials [77, 78]. Since hybrid films are suspected to have a high trap density and $E_{g,PbS-QD}$ of the applied PbS-QDs absorbing at 1240 nm was ~ 1 eV, a generation energy for electron hole pairs of approximately 4 eV was assumed corresponding to 250 electrons/keV for the conversion factor $\gamma_{conv,PbS}$. In figure 9.10 an offset level (545 fA) is observable, which is most likely assigned to measurement artifacts, since hybrid photodiodes for near-infrared did not show a measurable photoresponse without applied bias. Especially in thick films, generated charge carriers should recombine before reaching the electrodes. With the conversion factor γ_{conv} of 250 electrons/keV a maximum photocurrent of 15.9 pA could be possible, but only 1.9 pA (offset subtracted) can be extracted by the 2.7 µm device at a field of 37 Vµm^{-1} due to insufficient charge carrier mobilities and lifetimes yielding to short schubwegs. A interpolation of the response for 1000 µm layers reveals that although the X-ray absorption increases dramatically, a higher photocurrent cannot be extracted without improving the film properties with respect to the $\mu\tau$ product (see figure 9.11). 97 Vµm^{-1} of electric field needs to be applied with oBHJ conditions to withdraw all generated charge carriers from the device. Especially for improving charge carrier transport an optimization of the ligands need to be considered. A summary of the findings for these three different layers (2.7 µm and 1000 µm with different material properties) to compare the figures of the devices is given in table 9.3. Further work on these materials need to be performed to evaluate the applicability of this approach for X-ray detection.

9.2.3. Comparison of Quasi-Direct and Direct Conversion Devices

To summarize, a comparison of the response to X-ray exposure for both approaches of quasi-direct and direct conversion is given. In addition, references of thick oBHJ films (100 µm) and a standard OPD with an attached scintillator layer is included for comparison. All devices have the same active area

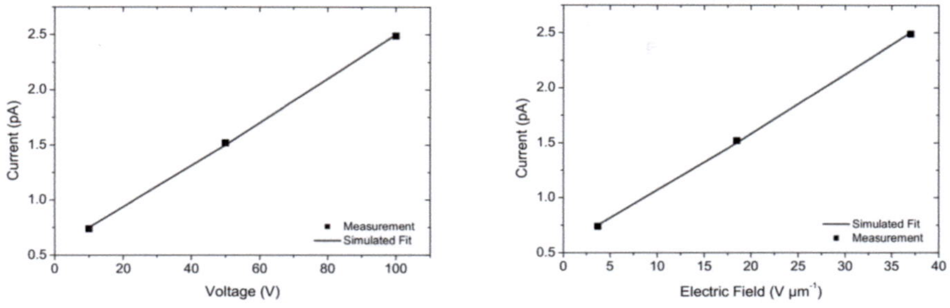

(a) Applied voltage (b) Electric Field accross the active layer

Figure 9.10.: Electronic response to a modulated X-ray beam at 50 kV and a dose of $\sim$ 200 µGy/s of a 2.7 µm thick hybrid photodiode with embedded PbS-QDs. The measured photoresponse of the device at the lock-in (black squares) and the fit by equation 9.1 and 9.2 with $\mu = 1.1 \times 10^{-5}$ cm^2(Vs)$^{-1}$, $\tau = 1.37$ µs ($\mu\tau = 1.5 \times 10^{-11}$ cm^2/V) and a conversion factor $\gamma_{conv,PbS}$ of 250 electrons/keV are plotted vs. the applied voltage (**a**) and the electric field (**b**). Note that an offset level of 545 fA was considered for fitting.

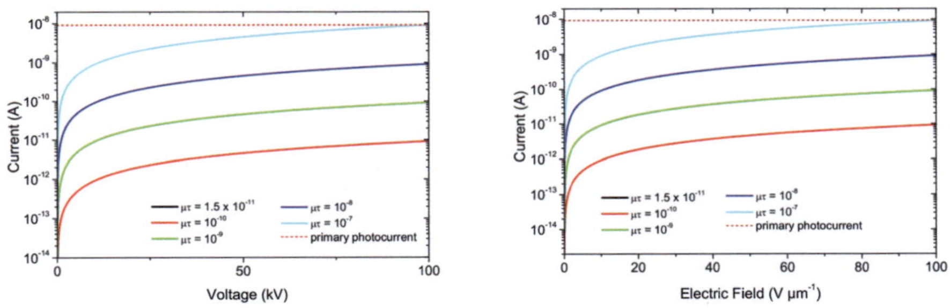

(a) Applied voltage (b) Electric Field accross the active layer

Figure 9.11.: Simulated response of a 1000 µm thick hybrid photodiode with embedded PbS-QDs to X-ray exposure of $\sim$ 200 µGy/s at 50 kV by the model of equation 9.1 and 9.2. The $\mu\tau$ product was varied from the obtained value of the measurement of 1.5×10^{-11} cm^2/V to an optimistic situation of oBHJs of 1.0×10^{-7} cm^2/V. The conversion efficiency $\gamma_{conv,PbS}$ was kept constant at 250 electrons/keV. In optimistic condition all of the photocurrent can be extracted for a 1000 µm layer at the expense of high voltages of 97 kV ((**a**)) and electric fields of 97 Vµm^{-1} ((**b**)). Note that no offset level of was considered for simulation.

Layer thickness	Photoel. absorption	Absorption. energy	Photon output	μ ($cm^2(Vs)^{-1}$)	τ (µs)	Primary photocurr.	Response	Elec. field
2.7 µm	0.64 %	5.9×10^5 keV	250	1.6×10^{-5}	1.37	15.9 pA	1.9 pA	37 $Vµm^{-1}$
1000 µm	83.15 %	8.3×10^7 keV	250	1.6×10^{-5}	1.37	13.3 nA	0.8 pA	37 $Vµm^{-1}$
1000 µm	83.15 %	8.3×10^7 keV	250	1.0×10^{-3}	100	13.3 nA	13.3 nA	97 $Vµm^{-1}$

Table 9.3.: X-ray response for the different layer thicknesses of hybrid PbS-QD devices are given with the corresponding parameters. The device with 2.7 µm layer thickness are values obtained from the fit, whereas for the 1000 µm layer the values were projected with the model of 9.1 for the current and for the optimistic situation with parameters from oBHJ films. Note that the offset level of 545 fA for the 2.7 µm device was subtracted.

of 4 mm^2. Considering the response only (see figure 9.12), the OPD reference shows the best signal with a scintillator of ~ 1 mm thickness and thus high X-ray absorption and subsequent fluorescence. The drop casted 100 µm oBHJ device shows only little signal at high voltages. This measurement and open circuit measurements were made to ensure that no parasitic signals were recorded. In contrast, the sample with Gd_2O_2S particles gives a increasing signal with the applied bias up to 900 V, although the scintillation losses due to milling were not taken into account. Hybrid devices with PbS-QDs show a similar behavior and need voltages of 100 V to have a significant response.

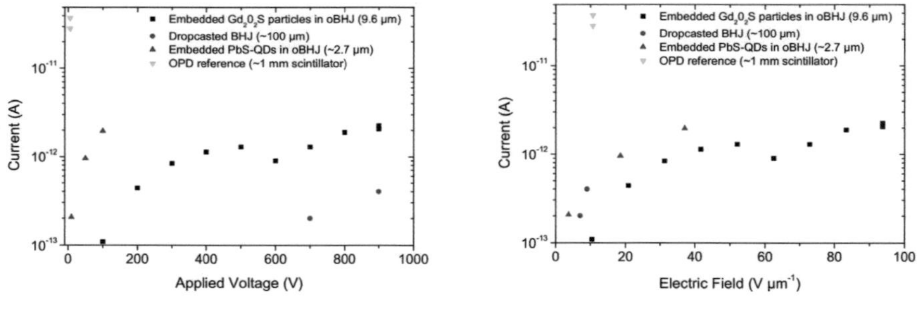

(a) Applied voltage

(b) Electric field across the active layer

Figure 9.12.: Electronic response to a modulated X-ray beam at 50 kV and an incident dose of ~ 200 µGy/s. Embedded scintillating particles of Gd_2O_2S in a oBHJ for quasi-direct conversion (black squares) as well as the embedded PbS-QDs in oBHJ for direct conversion (blue triangles) show a significant response while the drop casted oBHJ (red circles) shows a very low and noisy signal. As reference, a conventional OPD (green triangles) with an attached scintillator component (CsI) was measured, too. Note that Gd_2O_2S device were not corrected with the conversion loss after milling.

For further comparison the modulated response is normalized to the incident dose and the thickness of the active layer as presented in figure 9.13. Note that the response of the detector is normalized to the thickness of the attached scintillator or the effectively embedded conversion particles. With these conditions, the results of the approaches can be classified more realistically and a more reliable conclusion on the potential of Gd_2O_2S and PbS-QD in hybrid devices can be given.

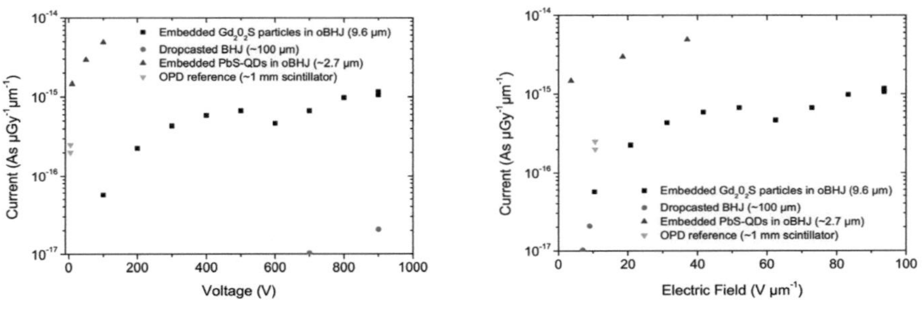

(a) Applied voltage

(b) Electric field across the active layer

Figure 9.13.: Electronic response to a modulated X-ray beam at 50 kV normalized to the thickness of the absorbing layer and the incident dose of $\sim$ 200 µGy/s. Embedded scintillating particles of Gd_2O_2S in a oBHJ for quasi-direct conversion (black squares) as well as the embedded PbS-QDs in oBHJ for direct conversion (blue triangles) show a significant response while the drop casted oBHJ (red circles) shows a very low and noisy signal. As reference a conventional OPD (green triangles) with an additional topping scintillator film (CsI) was measured, too. Note that Gd_2O_2S device were not corrected with the conversion loss after milling.

Accordingly, the results with PbS-QDs or Gd_2O_2S embedded particles in oBHJ show a truly possible alternative for X-ray detection. However, further investigations and optimizations are needed for efficient charge carrier extraction and thus high quantum yield for thick layer devices in future.

10. Outlook

In this work a wide range of promising properties and characteristics of hybrid composites for imaging and photodetection have been demonstrated. Especially, from a material's perspective future effort is required. PbS:P3HT:PCBM composites still suffer from fluctuations in material quality. The very sensitive material system needs components which do not show different properties according to different synthesis and manufacturers. In particular, for colloidal PbS-QDs a process needs to be introduced on industrial scale to ensure invariable, constant material properties from synthesis for high yields and reliability of hybrid devices. Since this work concentrated strongly on PbS:P3HT:PCBM and PCPDTBT for NIR detection, an extension of hybrid material systems to different QDs and polymers seems reasonable. Further materials provide more choice for suitable composites according to the application and a better coverage of the visible and infrared spectra.

Integration of IO-HPDs and OPDs for application in imaging will be a next important step to make. State-of-the-art CMOS-imaging devices are highly integrated on a single chip. Hybrid and organic material systems for NIR-imaging need to demonstrate and have to be developed for high integration. Compatibility to conventional systems is a major requirement to enter markets and applications more easily.

Besides the integration of hybrid and organic photodetectors to a-Si flat panels, a first proof of concept for the integration of OPDs to CMOS-imaging chips can be given and shall encourage future activities. Figure 10.1(a) shows a conventional chip for CMOS-imaging in mobile phones. A layout shows the dimensions of the reticle in figure 10.1(b). With integrating OPDs on the top of the CMOS chip as illustrated in figure 10.1(c), compatibility to conventional imaging applications can be achieved. This approach is realized with processing OPDs on top of the CMOS chip (see figure 10.1(d)) and a test image was captured from these OPD/CMOS imaging devices. Meanwhile, a similar approach showed also very promising results for the integration of IO-HPDs to CMOS-imaging devices [3].

Promising results for direct conversion of x-rays have to be reproduced and investigated. Further improvements and engineering of material and device properties to enhance sensitivity to x-rays will be a major task. However, direct

(a) Conventional chip for CMOS-imaging

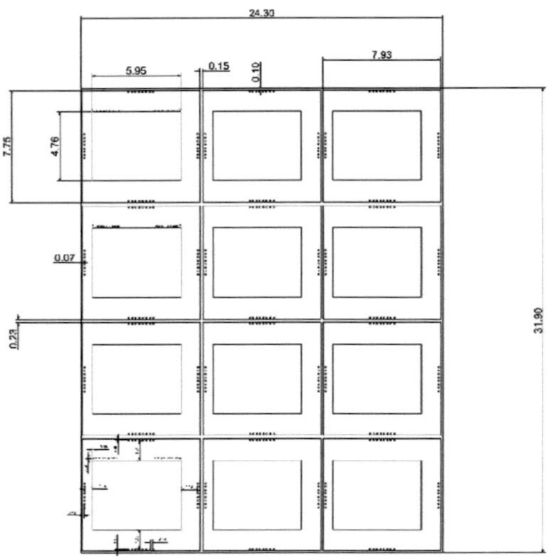

(b) Dimensions

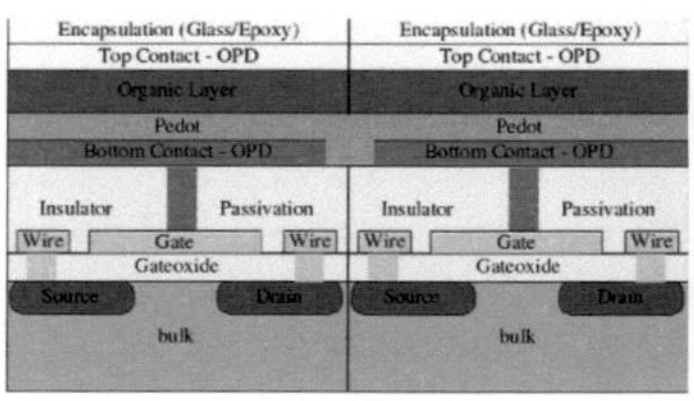

(c) Device stack

(d) OPD integrated on top of a CMOS imaging sensor

(e) Captured image

Figure 10.1.: Integ ration of OPDs on CMOS chips: (**a**) A CMOS chip with 9 dies for imaging. The two left upper ones are used for integration with OPDs. (**b**) The dimensions for the chip ($24.3\,\text{mm}^2 \times 31.9\,\text{mm}^2$) and with an active area of $5.95\,\text{mm}^2 \times 4.76\,\text{mm}^2$. (**c**). OPDs can be integrated on top of the CMOS chip utilizing a entire level for the photodetectors (3D integration). (**d**) A processed CMOS chip with OPDs on top with (**e**) an acquired image from the hybrid imaging chip.

conversion is of great interests for applications in x-ray imaging and detection to get rid of additional scintillating layers and materials. Competitive direct x-ray conversion hybrid devices can achieve low cost, processability and higher resolution capabilities for X-ray imaging.

This work shows a major future potential for organic and hybrid devices as modification of, e.g., the spectral response of organic photodiodes can be obtained by engineering and embedding additional materials (e.g. PbS-QDs) to the organic blend and integrating into conventional systems (e.g. imagers).

Appendix

A. List of used Symbols and Abbrevations

A.1. Symbols

A.1.1. English Symbols

C_D	depletion layer capacitance
C_i	intrinsic capacitance
D_n	diffusion coefficient for electrons
D_p	diffusion coefficient for holes
$\vec{E}$	electric field
E	energy
E^*	energy level of the first excitonic state
E_C	edge of the conduction band
$E_{Coulomb}$	Coulomb charging energy
$E_{F,n}$	Fermi energy level in n-doped semiconductors
$E_{F,p}$	Fermi energy level in p-doped semiconductors
E_F	Fermi energy level
E_g	energy bandgap
E_i	intrinsic Fermi energy
E_V	edge of the valence band
f	frequency
G	generation rate for charge carriers
g_m	transconductance
$G_{n,rad}$	generation rate for electrons by radiation
G_n	generation rate for electrons
$G_{p,rad}$	generation rate for holes by radiation
G_p	generation rate for holes
G_{PC}	generation rate in a photoconductor
$Gain_{PC}$	amplification / gain in a photoconductor
$\vec{J}$	current density
$\vec{J}_{diff,n}$	diffusion current density of electrons
$\vec{J}_{diff,p}$	diffusion current density of holes

$\vec{J}_{diff}$	diffusion current density
$\vec{J}_{drift,n}$	drift current density of electrons
$\vec{J}_{drift,p}$	drift current density of holes
$\vec{J}_{drift}$	drift current density
$\vec{J}_n$	current density of electrons
$\vec{J}_p$	current density of holes
$\hat{i}$	noise current
I_D	total current in a diode
I_{dis}	dissociation current
I_{DS}	drain-source current in a field-effect transistor
I_{ext}	extraction current
I_F	flicker noise current
I_J	Johnson noise current
I_M	measured current
I_{max}	maximum current of diode under illumination
I_{PC}	photocurrent in a photoconductor
I_{ph}	primary photocurrent
I_R	measured reference current of a calibrated device
I_S	saturation current in a diode
I_S	saturation current in a diode
I_S	shot noise current
I_{SC}	short circuit current of a photodiode
J_S	saturation current
k	wavenumber
L	inductance
m	mass
m_e	mass of an electron
m_e^*	effective mass of an electron
m_h	mass of an hole
m_h^*	effective mass of an hole
n	electron density
n	ideality factor of diodes
N_A	acceptor density
N_C	density of states in the conduction band
N_{col}	number of collected charges
N_D	donor density
n_{int}	internal quantum efficiency
N_{Ph}	number of incident photons
N_t	total density of states
N_V	density of states in the valence band
p	hole density
P_{opt}	optical power

R	recombination rate for charge carriers
R_n	recombination rate for electrons
R_p	recombination rate for holes
R_P	shunt resistance of a photodiode
R_{PD}	Spectral responsivity of a photodetector
R_S	serial resistance of a photodiode
S_{X-ray}	spectrum of X-rays
t_{tr}	transit time
U	net recombination rate in devices
U_{b-b}	band to band recombination
U_n	net recombination rate for electrons
U_p	net recombination rate for holes
U_s	recombination at surfaces
U_t	trap assisted recombination
V	potential
V_{bi}	built-in potential
V_{DS}	drain-source voltage in a field-effect transistor
V_{GS}	gate-source voltage in a field-effect transistor
V_{OC}	open circuit voltage of a photodiode
V_{th}	threshold voltage
W_D	depletion layer width
Z	impedance

A.1.2. Greek Symbols

α	absorption coefficient
ϵ_0	permittivity of vacuum
ϵ_r	relative permittivity
ϵ_s	permittivity of a semiconductor
γ_{conv}	conversion efficiency of scintillator for x-rays
λ	wavelength
μ	mobility
μ_e	electron mobility
μ_h	hole mobility
ν	frequency of radiation
Φ	work function
$\Psi(\vec{r})$	time-independent wave-function
$\Psi(\vec{r},t)$	time-independent wave-function
ρ	space charge density
σ	conductivity
σ_e	electron conductivity
σ_h	hole conductivity

τ	charge carrier lifetime
τ_e	electron lifetime
τ_h	hole lifetime
υ_{cp}	permeation probability
$\vec{\upsilon_D}$	drift velocity

A.1.3. Physical Constants

c	speed of light
$h = 2\pi \cdot \hbar$	Planck constant
k_B	Boltzmann constant
q	elementary charge

A.2. Abbrevations

ADC	analog digital converter
Ag	Silver
Al	Aluminum
AM	active matrix
AO	atomic orbital
a-Si	amorphous silicon
Au	Gold
Ca	Calcium
CELIV	charge extraction by linear increasing voltage
CNT	carbon nanotube
CQD	colloidal quantum dot
D*	detectivity
DUT	device under test
EBL	electron blocking layer
EQE	external quantum efficiency
ETL	electron transport layer
FET	field effect transistor

FF	fill factor
FFT	fast Fourier transformation
HBL	hole blocking layer
HOMO	highest occupied molecular orbital
HTL	hole transport layer
IO-HPD	inorganic-organic hybrid photodiode
ITO	indium tin oxide
LED	light emitting diode
LiF	Lithium Fluoride
LUMO	lowest unoccupied molecular orbital
MEH-PPV	poly[2-methoxy-t-(2′-ethylhexyloxy)-p-phenylene vinylene
MO	molecular orbital
MOSFET	metal oxide semiconductor field effect transistor
MTF	modulation transfer function
NEP	noise equivalent power
NIR	near-infrared
OA	oleic acid
oBHJ	organic bulk heterojunction
OFET	organic field effect transistor
OLED	organic light emitting diode
OPD	organic photodetector
OPV	organic photovoltaic
P3HT	Poly(3-hexylthiophene)
PCBM	[6,6]-Phenyl-C61-butric-acid-methyl-ester
PCPDTBT	Poly[2,6-(4,4-bis-(2-ethylhexyl)-4H-cyclopenta[2,1-b;3,4-b′]dithiophene)-alt-4,7-(2, 1,3-benzothiadiazole)]
PEDOT	Poly(3,4-ethylenedioxythiophene)
PPV	poly(p-phenylene-vinylene)

PSS	Poly(styrenesulfonate)
QD	quantum dot
RFID	radio frequency identification
RH	relative humidity
ROIC	readout integrated circuit
SEM	scanning electron microscopy
TCO	transparent conductive oxide
TEM	transmission electron microscopy
TFB	Poly[(9,9-dioctylfluorenyl-2,7-diyl)-co-(4,4′-(N-(4-sec-butylphenyl))diphenylamine)]
TFT	thin film transistor

List of Figures

1.1 The spectrum of electromagnetic radiation 2
1.2 Activities and scope of the work 3

2.1 Energy band gaps of conductors, semiconductors and insulators 5
2.2 Solution of the Schrdinger equation in a quantum well with infinite barriers 7
2.3 The Kronig Penney model and its approximation 9
2.4 Energy band gaps of semiconductors with direct and indirect transitions 10
2.5 Fermi energy levels due to n- and p-doping 12
2.6 One dimensional diffusion and drift of charge carriers 13
2.7 Beer's law of absorption and recombination and generation processes in semiconductors 17
2.8 A photoconductor made of a semiconductor 18
2.9 Photoconductive gain mechanism 20
2.10 The pn-junction 21
2.11 The Schottky contacts 24
2.12 Semiconductor heterojunctions 25
2.13 Types of colloidal quantum dots 26
2.14 Energy profile in the different types of quantum dots 27
2.15 Discrete states, bandgap and absorption of quantum dots 28
2.16 Growth of colloidal quantum dots 30
2.17 Type of polymers 31
2.18 Hybridization form of atomic orbitals for carbon atoms 32
2.19 Bonding in ethylene molecules with sp^2-hybridized carbon atoms 33
2.20 Chemical structure of polyacetylene. 34
2.21 Energy bandgaps in conjugated polymers 35
2.22 Delocalization of p-orbitals in conjugated polymers 35
2.23 Formation and states of solitons 36
2.24 Mesomeric structures of polythiophenes 36
2.25 Types of polarons 37
2.26 Types and states of excitons 37

2.27 Charge transport in organic semiconductors by hopping along discrete states in the layer. 38
2.28 The organic bulk heterojunction 40
2.29 The working principle of an organic bulk heterojunction in an organic photodetector. 41
2.30 Typical I-V characteristics of an OPD. 43
2.31 Equivalent circuit of an OPD. 43

3.1 Measurement setups for I-V characteristics 48
3.2 A standard layout for 8 testdiodes of 2 cm × 2 cm. 49
3.3 Setup for spectral EQE measurements 51
3.4 Setup for spatially resolved photocurrents 52
3.5 Setup for optical pulsed measurements 53
3.6 Setup for high frequency optical pulsed measurements 54
3.7 Layout of testdiodes used for fast transient measurements reproduced from ref. [129]. 54
3.8 Setup for noise analysis 56
3.9 Setup for impedance spectroscopy measurements. The impedance analyzer is the model HP 4192A from Hewlett-Packard is used. 58
3.10 Schematic for the read out electronics for imaging devices. . . . 59
3.11 Hardware realization of an ROIC for capturing images from an a-Si AM-TFT panel. 59
3.12 Flat-Field Correction 60
3.13 Modulated transfer function 61
3.14 Modulated transfer function of real imaging systems 62
3.15 Setup for accelerated lifetime measurements 63
3.16 Scanning and transmission electron microscopy. 64
3.17 The charge extraction by linear increasing voltage (CELIV) method 66
3.18 Setup of a CELIV-measurement 67
3.19 Transfer characteristics and determination of parameters 68
3.20 Layout of backgated field effect transistors 69

4.1 Used organic polymers and molecules 73
4.2 Schematic drawing of an oleic acid capped lead sulfide quantum dot 74
4.3 Chemical structure of (**a**) PEDOT:PSS and (**b**) TFB 75
4.4 Transmission of the commercial ITO substrate in the Vis-NIR spectra. 76
4.5 Deposition technique of spin-coating 78
4.6 Doctor-blading technique for film deposition from solution . . . 78
4.7 Deposition technique of spray-coating 79

4.8 Lithographic structuring of layers with positive and negative photoresists 80
4.9 Lift-off process with a positive photoresist 81
4.10 Thermal evaporation in an ultra-high vacuum chamber. 82
4.11 Process of sputtering in ultra-high vacuum. 83
4.12 Schematic layout of an organic and hybrid photodiode 84

5.1 I-V characteristics and EQE of doctor-bladed P3HT:PCBM photodetectors 88
5.2 Selected polymers with NIR sensitivity from literature 89
5.3 I-V characteristics and dark currents of different ratios of PCPDTBT and PCBM 90
5.4 Spectral sensitivity of PCPDTBT:PCBM with different PCBM contents 91
5.5 Post annealing effect on spray coated and doctor-bladed devices 93
5.6 Spectral response of pure P3HT:PCBM and PCPDTBT:PCBM blends 94
5.7 Combinations of P3HT:PCBM and PCPDTBT:PCBM blends for spectral absorption modifications 95
5.8 Light currents and EQE of stacked devices 96
5.9 I-V characteristics of P3HT:PCPDTBT:PCBM devices with varying composition 97
5.10 Spectral Response of P3HT:PCPDTBT:PCBM devices with varying content concentrations 98
5.11 Near-infrared light current densities and spectral response according to the PCPDTBT weight content 99
5.12 Absorption properties of P3HT and PCPDTBT 100
5.13 Fits of EQE for different P3HT:PCPDTBT:PCBM ratios 102
5.14 Noise current spectrum for a spray-coated PCPDTBT:PCBM device 103
5.15 Dynamic behavior of a PCPDTBT:PCBM photodiode with an active layer thickness of 550 nm 104
5.16 Dynamic behavior of a PCPDTBT:PCBM photodiode with an active layer thickness of 800 nm 105

6.1 Schematic illustration of a hybrid photodiode with solution processable QDs 108
6.2 Flat band energy diagram of PbS-QDs:P3HT:PCBM device . . . 110
6.3 I-V characteristics of PbS-QDs:P3HT, PbS-QDs:PCBM and PbS-QDs:P3HT:PCBM devices 111
6.4 Distributions of dark and light currents for dual and ternary blends 112
6.5 Spectral response of devices with dual and ternary blends 112

6.6 Absorption spectra of P3HT, PCBM and PbS-QDs films 113
6.7 I-V characteristics and spectral responses of IO-HPDs with increasing PbS-QD content 114
6.8 I-V characteristics of hybrid photodiodes with TFB as interlayer 116
6.9 Spectral response of hybrid photodiodes with TFB / PEDOT:PSS or O_2 plasma treated TFB interlayer 117
6.10 I-V characteristics and EQEs of devices with self-assembled layers of PSS 118
6.11 I-V characteristics and on-off ratios of IO-HPDs with top contacts of different work functions 119
6.12 Spectral sensitivity of hybrid devices with different top contacts 119
6.13 Dark and photosensitivity according to the top electrodes' work function 120
6.14 I-V characteristics of spray-coated IO-HPDs of different thicknesses 121
6.15 Spectral response and field-dependence of a 970 nm and a 2700 nm spray-coated device with reference to a 200 nm doctor-bladed device 122
6.16 Tunability of the spectral response corresponding to QD-diameter 123
6.17 Spectral response of IO-HPDs with a mixture of three different PbS-QDs 124
6.18 I-V characteristics and spectral response of IO-HPDs with butylamine capped PbS-QDs 125
6.19 I-V characteristics of IO-HPDs with oxidized PbS-QD batches . . 126
6.20 Spectral response of IO-HPDs with oxidized PbS-QD batches . . 127
6.21 Voltage dependence of the spectral response for IO-HPDs with oxidized PbS-QD batches 127
6.22 I-V characteristics and spectral sensitivity of a hybrid state of the art device 129

7.1 SEM images of bilayer devices of PbS-QD:P3HT 132
7.2 SEM images of a doctor-bladed and a spray-coated film of PbS-QD:P3HT:PCBM 133
7.3 TEM images of a doctor-bladed PbS-QD:P3HT:PCBM film 133
7.4 FET transfer characteristics for hybrid organic/inorganic composites 136
7.5 Electron and hole mobilities in different hybrid organic/inorganic composites. 137
7.6 Dark and NIR Photo-CELIV traces for an IO-HPD 138
7.7 Extracted mobilities by dark CELIV and Photo-CELIV measurements at 1310 nm 140

7.8 Extracted CELIV mobilities in dependence on the slope of voltage ramp . . . 140
7.9 Photoluminescence spectra of PbS-QD:P3HT:PCBM devices . . . 141
7.10 Spatial resolved photocurrents of IO-HPDs with excitation at 532 nm and 940 nm . . . 143
7.11 Normalized 3D plot for spatial resolved photocurrents of IO-HPDs with excitation at 532 nm and 940 nm . . . 143
7.12 I-V characteristics of an IO-HPD with a temperature sweep from 0 °C to 80 °C . . . 144
7.13 Temperature dependence of saturation and dark currents of IO-HPDs . . . 146
7.14 Characteristics of NIR photoresponse of IO-HPDs . . . 147
7.15 I-V characteristics and the resulting fit of an IO-HPD with PbS-QD absorbing at 1220 nm by the introduced modified model. . . . 148
7.16 I-V characteristics for the devices used for fast pulsed electro-optical measurements . . . 149
7.17 Temporal response for an IO-HPD excited at 532 nm by a laser pulse of 2.5 ns in dependence of the applied voltage . . . 150
7.18 Temporal response for an IO-HPD excited at 532 nm by a laser pulse of 2.5 ns in dependence of the incident power . . . 150
7.19 Fit of the temporal response of IO-HPDs for optical intensities of 928 µW . . . 151
7.20 Fit of the temporal response of IO-HPDs for optical intensities of 0.7 µW . . . 152
7.21 Equivalent circuit used for the description of impedance spectroscopy measurements. . . . 153
7.22 Obtained overall resistances R(f) and capacitances C(f) by impedance spectroscopy . . . 154
7.23 Impedance spectroscopy measurements at an applied bias of 0.8 V with fits for R(f) and C(f) . . . 154
7.24 Impedance spectroscopy measurements at an applied bias of 0.0 V with fits for R(f) and C(f) . . . 155
7.25 Noise over frequency for different applied reverse voltages. . . . 156
7.26 Noise current densities at 170 Hz (**a**) for different voltages and (**b**) for different dark current . . . 156
7.27 Accelerated lifetime and shelf lifetime measurements for IO-HPDs . . . 157

8.1 Schematic illustration of the imager with an a-Si AM backplane and an unstructured inorganic/organic hybrid frontplane . . . 160

8.2 NIR response of an imager with a PbS-QD sensitized IO-HPD in comparison to an imager with a P3HT:PCBM OPD . 162
8.3 Near-infrared shadow cast of a transparent slide showing a monarch butterfly . . . 163
8.4 Crosstalk analysis of an IO-HPD . . . 164
8.5 MTF of the imager to define resolution properties . . . 165
8.6 Snapshots from a NIR movie acquired by an IO-HPD imager . . 166

9.1 Indirect and direct conversion of X-rays . . . 168
9.2 Stack of an a-Si flat detector for X-ray imaging . . . 169
9.3 Layout of an a-Se flat detector . . . 170
9.4 Direct and quasi-direct conversion in IO-HPDs . . . 171
9.5 The mass-attenuation factor for the photoelectric absorption of Gd_2O_2S and PbS . . . 172
9.6 I-V characteristics of a hybrid device with scintillator particles . 174
9.7 X-ray response of hybrid photodiodes with embedded Gd_2O_2S particles . . . 175
9.8 Interpolating of Gd_2O_2S devices for a layer thickness of 1000 µm from measurements . . . 176
9.9 TEM image and I-V characteristics of a 2.7 µm device under dark and illuminated conditions . . . 179
9.10 Electronic response to a modulated X-ray beam at 50 kV and a dose of $\sim$ 200 µGy/s of a 2.7 µm thick hybrid photodiode with embedded PbS-QDs . . . 181
9.11 Simulated response of a 1000 µm thick hybrid photodiode with embedded PbS-QDs to X-ray exposure of $\sim$ 200 µGy/s at 50 kV . . . 181
9.12 Electronic response to a modulated X-ray beam at 50 kV and an incident dose of $\sim$ 200 µGy/s for investigated devices . 182
9.13 Electronic response to a modulated X-ray beam at 50 kV normalized to the thickness of the absorbing layer and the incident dose of $\sim$ 200 µGy/s for investigated devices . . . 183

10.1 Integration of OPDs on CMOS chips . . . 186

List of Tables

3.1 Different types of illumination sources applied to measurements of I-V characteristics. 49

5.1 Potential near-infrared sensitive polythiophene derivatives with reported characteristics. 88
5.2 Dark currents and serial resistances in dependence of thermal post-annealing 93
5.3 Blending ratios, fractions and EQEs at 515 nm and 725 nm of P3HT:PCBM and PCPDTBT:PCBM films 96
5.4 Derived parameters for the different blends of PCPDTBT:P3HT:PCBM for film thicknesses and extraction efficiencies 100

6.1 Relation of dark currents and EQEs according to PbS-QD weight contents 114

7.1 Analyzed composites in OFETs and the measured mobilities with corresponding voltages V_{GS} and V_{DS}. 135
7.2 Parameters used for equation 7.8 to fit the exponential decay of the temporal response of IO-HPDs. 152
7.3 Parameters used for equation 7.8 to fit the exponential decay of the temporal response of IO-HPDs. 154

9.1 Relevant properties of applied materials for quasi-direct (Gd_2O_2S) and direct conversion (PbS) of X-rays in hybrid devices 172
9.2 X-ray response for the different layer thicknesses of Gd_2O_2S devices 177
9.3 X-ray response for the different layer thicknesses of hybrid PbS-QD devices 182

Bibliography

[1] P. M. Allemand, A. Koch, F. Wudl, Y. Rubin, F. Diederich, M. M. Alvarez, S. J. Anz, and R. L. Whetten. Two different fullerenes have the same cyclic voltammetry. *Journal of the American Chemical Society*, 113(3):1050–1051, 1991.

[2] T. D. Anthopoulos, C. Tanase, S. Setayesh, E. J. Meijer, J. C. Hummelen, P. W. M. Blom, and D. M. de Leeuw. Ambipolar organic field-effect transistors based on a solution-processed methanofullerene. *Adv. Mater.*, 16(23-24):2174–2179, 2004.

[3] D. Baierl, L. Pancheri, M. Schmidt, D. Stoppa, G.-F. Dalla Betta, G. Scarpa, and P. Lugli. A hybrid cmos-imager with a solution-processable polymer as photoactive layer. *Nat Commun*, 3:1175–, Nov. 2012.

[4] Z. Bao, A. Dodabalapur, and A. J. Lovinger. Soluble and processable regioregular poly(3-hexylthiophene) for thin film field-effect transistor applications with high mobility. *Appl. Phys. Lett.*, 69:4108–4110, 1996.

[5] D. A. R. Barkhouse, R. Debnath, I. J. Kramer, D. Zhitomirsky, A. G. Pattantyus-Abraham, L. Levina, L. Etgar, M. Grätzel, and E. H. Sargent. Depleted bulk heterojunction colloidal quantum dot photovoltaics. *Advanced Materials*, 23(28):3134–3138, 2011.

[6] J. B. Barton, R. F. Cannata, and S. M. Petronio. Ingaas nir focal plane arrays for imaging and dwdm applications. *Proc. SPIE*, 37:4721, 2002.

[7] H. Bässler. Localized states and electronic transport in single component organic solids with diagonal disorder. *physica status solidi (b)*, 107(1):9–54, 1981.

[8] H. Bässler. Charge transport in disordered organic photoconductors a monte carlo simulation study. *physica status solidi (b)*, 175(1):15–56, 1993.

[9] M. Böberl, M. Kovalenko, S. Gamerith, E. J. W. List, and W. Heiss. Inkjet-printed nanocrystal photodetectors operating up to 3um wavelengths. *Advanced Materials*, 19(21):3574–3578, 2007.

[10] M. Böberl, M. V. Kovalenko, G. Pillwein, G. Brunthaler, and W. Heiss. Quantum dot nanocolumn photodetectors for light detection in the infrared. *Applied Physics Letters*, 92(26):261113, 2008.

[11] C. J. Brabec, A. Cravino, D. Meissner, N. S. Sariciftci, T. Fromherz, M. T. Rispens, L. Sanchez, and J. C. Hummelen. Origin of the open circuit voltage of plastic solar cells. *Advanced Functional Materials*, 11(5):374–380, 2001.

[12] S. Braun, W. R. Salaneck, and M. Fahlman. Energy-level alignment at organic/metal and organic/organic interfaces. *Advanced Materials*, 21(14-15):1450–1472, 2009.

[13] L. Brus. Electronic wave functions in semiconductor clusters: experiment and theory. *The Journal of Physical Chemistry*, 90(12):2555–2560, 1986.

[14] J. H. Burroughes, D. D. C. Bradley, A. R. Brown, R. N. Marks, K. Mackay, R. H. Friend, P. L. Burns, and A. B. Holmes. Light-emitting diodes based on conjugated polymers. *Nature*, 347(6293):539–541, Oct. 1990.

[15] L. Cademartiri, E. Montanari, G. Calestani, A. Migliori, A. Guagliardi, and G. A. Ozin. Size-dependent extinction coefficients of pbs quantum dots. *J. Am. Chem. Soc.*, 128(31):10337–10346, Aug. 2006.

[16] H.-Y. Chen, L. K. F., G. Yang, H. G. Monbouquette, and Y. Yang. Nanoparticle-assisted high photoconductive gain in composites of polymer and fullerene. *Nat Nano*, 3(9):543–547, Sept. 2008.

[17] T.-A. Chen, X. Wu, and R. D. Rieke. Regiocontrolled synthesis of poly(3-alkylthiophenes) mediated by rieke zinc: Their characterization and solid-state properties. *Journal of the American Chemical Society*, 117(1):233–244, 1995.

[18] C. K. Chiang, C. R. Fincher, Y. W. Park, A. J. Heeger, H. Shirakawa, E. J. Louis, S. C. Gau, and A. G. MacDiarmid. Electrical conductivity in doped polyacetylene. *Phys. Rev. Lett.*, 39:1098–1101, 1977.

[19] Z. Chiguvare and V. Dyakonov. Trap-limited hole mobility in semiconducting poly(3-hexylthiophene). *Phys. Rev. B*, 70(23):235207, Dec 2004.

[20] D. Chirvase, Z. Chiguvare, M. Knipper, V. Parisi, J. Dyakonov, and J. Hummelen. Electrical and optical design and characterisation of regioregular poly(3-hexylthiophene-2,5diyl)/fullerene-based heterojunction polymer solar cells. *Synth. Met.*, 138:299–304, 2003.

[21] S. Choulis, V.-E. Choong, A. Patwardhan, M. Mathai, and F. So. Interface modification to improve hole-injection properties in organic electronic devices. *Advanced Functional Materials*, 16(8):1075–1080, 2006.

[22] S. A. Choulis, Y. Kim, J. Nelson, D. D. C. Bradley, M. Giles, M. Shkunov, and I. McCulloch. High ambipolar and balanced carrier mobility in regioregular poly(3-hexylthiophene). *Applied Physics Letters*, 85(17):3890–3892, 2004.

[23] L. Chu, A. Zrenner, G. Böhm, and G. Abstreiter. Normal-incident intersubband photocurrent spectroscopy on inas/gaas quantum dots. *Appl. Phys. Lett.*, 75:3599–4001, 1999.

[24] L.-L. Chua, J. Zaumseil, J.-F. Chang, E. C.-W. Ou, P. K.-H. Ho, H. Sirringhaus, and R. H. Friend. General observation of n-type field-effect behaviour in organic semiconductors. *Nature*, 434(7030):194–199, Mar. 2005.

[25] cintelliq ltd., editor. *256x256 pixels active matrix photo detector combing an amorphous Silicon thin film transistor array with organic photo diodes*, 2007.

[26] J. P. Clifford, G. Konstantatos, K. W. Johnston, S. Hoogland, L. Levina, and E. H. Sargent. Fast, sensitive and spectrally tuneable colloidal-quantum-dot photodetectors. *Nat. Nanotechnol.*, 4(1):40–44, Jan. 2009.

[27] Columbia University Clean Room. Spin coat theory. `http://www.clean.cise.columbia.edu/process/`, 2009.

[28] J. Cui, A. Wang, N. L. Edleman, J. Ni, P. Lee, N. R. Armstrong, and T. J. Marks. Indium tin oxide alternatives - high work function transparent conducting oxides as anodes for organic light-emitting diodes. *Adv. Mater.*, 13(19):1476–1480, 2001.

[29] J. N. de Freitas, I. R. Grova, L. C. Akcelrud, E. Arici, N. S. Sariciftci, and A. F. Nogueira. The effects of cdse incorporation into bulk heterojunction solar cells. *J. Mater. Chem.*, 20:4845–4853, 2010.

[30] C. de Mello Doneg, P. Liljeroth, and D. Vanmaekelbergh. Physicochemical evaluation of the hot-injection method, a synthesis route for monodisperse nanocrystals. *Small*, 1(12):1152–1162, 2005.

[31] R. Debnath, O. Bakr, and E. H. Sargent. Solution-processed colloidal quantum dot photovoltaics: A perspective. *Energy Environ. Sci.*, 4:4870–4881, 2011.

[32] C. Deibel and V. Dyakonov. Polymer-fullerene bulk heterojunction solar cells. *Reports on Progress in Physics*, 73(9):096401, 2010.

[33] C. D. Dimitrakopoulos and P. R. L. Malenfant. Organic thin film transistors for large area electronics. *Adv. Mater.*, 14:99–117, 2002.

[34] D. M. N. M. Dissanayake, R. A. Hatton, T. Lutz, R. J. Curry, and S. R. P. Silva. Charge transfer between acenes and pbs nanocrystals. *Nanotechnology*, 20(19):195205, 2009.

[35] R. F. Egerton, editor. *Physical Principles of Electron Microscopy*. Springer Science+Business Media, Inc, Boston, MA, 2005. Elektronische Ressource.

[36] M. M. Erwin, J. McBride, A. V. Kadavanich, and S. J. Rosenthal. Effects of impurities on the optical properties of poly-3-hexylthiophene thin films. *Thin Solid Films*, 409(2):198 – 205, 2002.

[37] N. I. for Standards and Technology. Physical reference data. http://www.nist.gov/physlab/data/index.cfm, 2010.

[38] M. A. Fox. Photoinduced electron transfer. *Photochemistry and Photobiology*, 52(3):617–627, 1990.

[39] A. Franceschetti, J. M. An, and A. Zunger. Impact ionization can explain carrier multiplication in pbse quantum dots. *Nano Lett.*, 6(10):2191–2195, Oct. 2006.

[40] R. Franke, B. Maennig, A. Petrich, and M. Pfeiffer. Long-term stability of tandem solar cells containing small organic molecules. *Solar Energy Materials and Solar Cells*, 92(7):732 – 735, 2008. Degradation and Stability of Polymer and Organic Solar Cells.

[41] S. V. Gaponenko. *Introduction to Nanophotonics*. Cambridge University Press, 2010.

[42] G. Garcia-Belmonte, A. Munar, E. M. Barea, J. Bisquert, I. Ugarte, and R. Pacios. Charge carrier mobility and lifetime of organic bulk heterojunctions analyzed by impedance spectroscopy. *Organic Electronics*, 9(5):847 – 851, 2008.

[43] W. Geens, T. Aernouts, J. Poortmans, and G. Hadziioannou. Organic co-evaporated films of a ppv-pentamer and c60: model systems for donor/acceptor polymer blends. *Thin Solid Films*, 403-404(0):438 – 443, 2002. ¡ce:title¿Proceedings of Symposium P on Thin Film Materials for Photovoltaics¡/ce:title¿.

[44] W. Geens, T. Martens, J. Poortmans, T. Aernouts, J. Manca, L. Lutsen, P. Heremans, S. Borghs, R. Mertens, and D. Vanderzande. Modelling the short-circuit current of polymer bulk heterojunction solar cells. *Thin Solid Films*, 451-452:498–502, Mar. 2004.

[45] G. E. Giakoumakis, C. D. Nomicos, E. N. Yiakoumakis, and E. K. Evangelou. Absolute efficiency of rare earth oxysulphide screens in reflection mode observation. *Physics in Medicine and Biology*, 35(7):1017–1023, 1990.

[46] W.-M. Gnehr. *Untersuchungen zum reaktiven Pulsmagnetronsputtern von ITO von metallischen Targets*. PhD thesis, Technischen Universität Ilmenau, 2006.

[47] A. Gocalinska, M. Saba, F. Quochi, M. Marceddu, K. Szendrei, J. Gao, M. A. Loi, M. Yarema, R. Seyrkammer, W. Heiss, A. Mura, and G. Bongiovanni. Size-dependent electron transfer from colloidal pbs nanocrystals to fullerene. *The Journal of Physical Chemistry Letters*, 1(7):1149–1154, 2010.

[48] C. Goh, R. J. Kline, M. D. McGehee, E. N. Kadnikova, and J. M. J. Fréchet. Molecular-weight-dependent mobilities in regioregular poly(3-hexyl-thiophene) diodes. *Applied Physics Letters*, 86(12):122110, 2005.

[49] X. Gong, M. Tong, Y. Xia, W. Cai, J. S. Moon, Y. Cao, G. Yu, C.-L. Shieh, B. Nilsson, and A. J. Heeger. High-Detectivity Polymer Photodetectors with Spectral Response from 300 nm to 1450 nm. *Science*, 325(5948):1665–1667, 2009.

[50] Goodfellow GmbH. `https://www.goodfellow.com/homee.aspx?LangType=1031`, 11 2010.

[51] L. Groenendaal, F. Jonas, D. Freitag, H. Pielartzik, and J. R. Reynolds. Poly(3,4-ethylenedioxythiophene) and its derivatives: Past, present, and future. *Adv. Mater.*, 12(7):481–494, 2000.

[52] I. Gur, N. A. Fromer, M. L. Geier, and A. P. Alivisatos. Air-stable all-inorganic nanocrystal solar cells processed from solution. *Science*, 310(5747):462–465, 2005.

[53] H. L. Harald Ibach. *Festkörperphysik: Einführung in die Grundlagen*. Springer Verlag, 5. auflage edition, 1991.

[54] H.C. Starck GmbH. `http://www.clevios.com/index.php?page_id=995&prod_service_id=325&anw_id=0&operate=&suchstart=938&prodselect_5=325|&suchstart=938&anwselect_14=0&suchfeld=`, 2010.

[55] M. A. Hines and G. D. Scholes. Colloidal pbs nanocrystals with size-tunable near-infrared emission: Observation of post-synthesis self-narrowing of the particle size distribution. *Advanced Materials*, 15(21):1844–1849, 2003.

[56] E. Holder, N. Tessler, and A. L. Rogach. Hybrid nanocomposite materials with organic and inorganic components for opto-electronic devices. *J. Mater. Chem.*, 18:1067–1078, 2008.

[57] H. Hoppe and N. Sariciftci. Polymer solar cells. In S. Marder and K.-S. Lee, editors, *Photoresponsive Polymers II*, volume 214 of *Advances in Polymer Science*, pages 1–86. Springer Berlin / Heidelberg, 2008.

[58] H. Hoppe and N. S. Sariciftci. Organic solar cells: An overview. *J. Mat. Res.*, 19:1924–1945, 2004.

[59] H. Hoppe and N. S. Sariciftci. Morphology of polymer/fullerene bulk-heterojunction solar cells. *J. Mater. Chem.*, 16:45–61, 2006.

[60] G. Horowitz. Organic field-effect transistors. *Adv. Mater.*, 10:365–377, 1998.

[61] H. Huang, A. Dorn, G. P. Nair, V. Bulovic, and M. G. Bawendi. Bias-induced photoluminescence quenching of single colloidal quantum dots embedded in organic semiconductors. *Nano Letters*, 7(12):3781–3786, Dec. 2007.

[62] S. Hunklinger. *Festkörperphysik*. Oldenbourg, München, 2., verb. aufl. edition, 2009. In: Oldenbourg-link.com. - Elektronische Ressource.

[63] I.-W. Hwang, C. Soci, D. Moses, Z. Zhu, D. Waller, R. Gaudiana, C. J. Brabec, and A. J. Heeger. Ultrafast electron transfer and decay dynamics in a small band gap bulk heterojunction material. *Adv. Mater.*, 19:2307–2312, 2007.

[64] B.-R. Hyun, Y.-W. Zhong, A. C. Bartnik, L. Sun, H. D. Abruna, F. W. Wise, J. D. Goodreau, J. R. Matthews, T. M. Leslie, and N. F. Borrelli. Electron injection from colloidal pbs quantum dots into titanium dioxide nanoparticles. *ACS Nano*, 2(11):2206–2212, 2008.

[65] R. M. Inc. `http://www.riekemetals.com/material_scinence_division.html`, 2010.

[66] A. Intaniwet, C. A. Mills, M. Shkunov, H. Thiem, J. L. Keddie, and P. J. Sellin. Characterization of thick film poly(triarylamine) semiconductor diodes for direct x-ray detection. *Journal of Applied Physics*, 106(6):064513, 2009.

[67] G. Itskos, A. Othonos, T. Rauch, S. F. Tedde, O. Hayden, M. V. Kovalenko, W. Heiss, and S. A. Choulis. Optical properties of organic semiconductor blends with near-infrared quantum-dot sensitizers for light harvesting applications. *Advanced Energy Materials*, 1(5):802–812, 2011.

[68] D. Jarzab, K. Szendrei, M. Yarema, S. Pichler, W. Heiss, and M. A. Loi. Charge-separation dynamics in inorganic-organic ternary blends for efficient infrared photodiodes. *Advanced Functional Materials*, 21(11):1988–1992, 2011.

[69] K. S. Jeong, J. Tang, H. Liu, J. Kim, A. W. Schaefer, K. Kemp, L. Levina, X. Wang, S. Hoogland, R. Debnath, L. Brzozowski, E. H. Sargent, and J. B. Asbury. Enhanced mobility-lifetime products in pbs colloidal quantum dot photovoltaics. *ACS Nano*, 0(0):null, 0.

[70] G. Juka, K. Geneviius, K. Arlauskas, R. sterbacka, and H. Stubb. Charge transport at low electric fields in π-conjugated polymers. *Phys. Rev. B*, 65(23):233208, Jun 2002.

[71] G. Juska, K. Arlauskas, and M. Viliunas. Extraction current transients: New method of study of charge transport in microcrystalline silicon. *Phys. Rev. Lett.*, 84(21):4946–4949, May 2000.

[72] G. Juska, K. Arlauskas, M. Viliunas, M.finas, K. Genevius, R. Österbacka, and H. Stubb. Charge transport in π-conjugated polymers from extraction current transients. *Phys. Rev. B*, 62(24):R16235–R16238, 2000.

[73] M. Kabir, S. Kasap, W. Zhao, and J. Rowlands. Direct conversion x-ray sensors: sensitivity, dqe and mtf. *Circuits, Devices and Systems, IEE Proceedings -*, 150:258–266, 2003.

[74] J.-E. Kallhammer. Imaging: The road ahead for car night-vision. *Nat Photon*, sample(sample):12–13, Sept. 2006.

[75] I. Kandarakis, D. Cavouras, G. S. Panayiotakis, and C. D. Nomicos. Evaluating x-ray detectors for radiographic applications: A comparison of znscds:ag with and screens. *Physics in Medicine and Biology*, 42(7):1351–1373, 1997.

[76] K. Kaneto, K. Hatae, S. Nagamatsu, W. Takashima, S. S. Pandey, K. Endo, and M. Rikukawa. Photocarrier mobility in regioregular poly(3-hexylthiophene) studied by the time of flight method. *Japanese Journal of Applied Physics*, 38(Part 2, No. 10B):L1188–L1190, 1999.

[77] S. O. Kasap and J. A. Rowlands. Review x-ray photoconductors and stabilized a-se for direct conversion digital flat-panel x-ray image-detectors. *J. Mater. Sci.: Mater. Electron.*, 11(3):179–198, Apr. 2000.

[78] S. O. Kasap and J. A. Rowlands. Direct-conversion flat-panel x-ray image sensors for digital radiography. *PROCEEDINGS OF THE IEEE,*, 90:591–604, 2002.

[79] H. E. Katz and J. Huang. Thin-film organic electronic devices. *ANNUAL REVIEW OF MATERIALS RESEARCH*, 39:71–92, 2009.

[80] J. Y. Kim, K. Lee, N. E. Coates, D. Moses, T.-Q. Nguyen, M. Dante, and A. J. Heeger. Efficient tandem polymer solar cells fabricated by all-solution processing. *Science*, 317:222–225, 2007.

[81] S. J. Kim, W. J. Kim, Y. Sahoo, A. N. Cartwrigh, and P. N. Prasad. Multiple exciton generation and electrical extraction from a pbse quantum dot photoconductor. *Appl. Phys. Lett.*, 92:031107, 2008.

[82] Y. Kim, S. A. Choulis, J. Nelson, D. D. C. Bradley, S. Cook, and J. R. Durrant. Device annealing effect in organic solar cells with blends of regioregular poly(3-hexylthiophene) and soluble fullerene. *Applied Physics Letters*, 86(6):063502–+, Feb. 2005.

[83] S. Kirchmeyer and K. Reuter. Scientific importance, properties and growing applications of poly(3,4-ethylenedioxythiophene). *J. Mater. Chem.*, 15:2077–2088, 2005.

[84] E. J. D. Klem, H. Shukla, S. Hinds, D. D. MacNeil, L. Levina, and E. H. Sargent. Impact of dithiol treatment and air annealing on the conductivity, mobility, and hole density in pbs colloidal quantum dot solids. *Applied Physics Letters*, 92(21):212105, 2008.

[85] G. I. Koleilat, L. Levina, H. Shukla, S. H. Myrskog, S. Hinds, A. G. Pattantyus-Abraham, and E. H. Sargent. Efficient, stable infrared photovoltaics based on solution-cast colloidal quantum dots. *ACS Nano*, 2(5):833–840, May 2008.

[86] G. Konstantatos, I. Howard, A. Fischer, S. Hoogland, J. Clifford, E. Klem, L. Levina, and E. H. Sargent. Ultrasensitive solution-cast quantum dot photodetectors. *Nature*, 442(7099):180–183, July 2006.

[87] G. Konstantatos and E. H. Sargent. Nanostructured materials for photon detection. *Nat Nano*, 5(6):391–400, June 2010.

[88] G. Konstantatos and E. H. Sargent. Colloidal quantum dot photodetectors. *Infrared Physics & Technology*, 54(3):278 – 282, 2011. ¡ce:title¿Proceedings of the International Conference on Quantum Structure Infrared Photodetector (QSIP) 2010¡/ce:title¿.

[89] M. V. Kovalenko, E. Kaufmann, D. Pachinger, J. Roither, M. Huber, J. Stangl, G. Hesser, F. Schäffler, and W. Heiss. Colloidal hgte nanocrystals with widely tunable narrow band gap energies: From telecommunications to molecular vibrations. *Journal of the American Chemical Society*, 128(11):3516–3517, 2006. PMID: 16536514.

[90] M. Kuno, K. A. Higginson, S. B. Qadri, M. Yousuf, S. H. Lee, B. L. Davis, and H. Mattoussi. Molecular clusters of binary and ternary mercury chalcogenides: Colloidal synthesis, characterization, and optical spectra. *The Journal of Physical Chemistry B*, 107(24):5758–5767, 2003.

[91] E. Kymakis and G. A. J. Amaratunga. Optical properties of polymer-nanotube composites. *Synthetic Metals*, 142(1-3):161 – 167, 2004.

[92] E. Kymakis, E. Koudoumas, I. Franghiadakis, and G. A. J. Amaratunga. Post-fabrication annealing effects in polymer-nanotube photovoltaic cells. *Journal of Physics D: Applied Physics*, 39(6):1058, 2006.

[93] C. A. Leatherdale, C. R. Kagan, N. Y. Morgan, S. A. Empedocles, M. A. Kastner, and M. G. Bawendi. Photoconductivity in cdse quantum dot solids. *Phys. Rev. B*, 62(4):2669–2680, Jul 2000.

[94] B. Lee, P. Podsiadlo, S. Rupich, D. V. Talapin, T. Rajh, and E. V. Shevchenko. Comparison of structural behavior of nanocrystals in randomly packed films and long-range ordered superlattices by time-resolved small angle x-ray scattering. *J. Am. Chem. Soc.*, 131(45):16386–16388, Nov. 2009.

[95] D. R. Lide and W. M. Hayne. *CRC Handbook of Chemistry and Physics*. CRC Press, 90th ed. edition, 2009–2010.

[96] H. C. Liu, L. Li, M. B. andZ. R. Wasilewski, G. J. Brown, F. Szmulowicz, and S. M. Hegde. A study of gaas/algaas p-type quantum well infrared photodetectors with different barrier heights. *J. Appl. Phys.*, 83:585–587, 1998.

[97] H. C. Liu, F. Szmulowicz, Z. R. Wasilewski, M. Buchanan, and G. J. Brown. Intersubband infrared detector with optimized valence band quantum wells for 3 to 5 mm wavelength region. *J. Appl. Phys.*, 85:2972–2976, 1999.

[98] J. Lorrmann, B. H. Badada, O. Inganäs, V. Dyakonov, and C. Deibel. Charge carrier extraction by linearly increasing voltage: Analytic framework and ambipolar transients. *Journal of Applied Physics*, 108(11):113705, 2010.

[99] G. W. Ludwig. X-ray efficiency of powder phosphors. *J. Electrochem. Soc.*, 118(7):1152–1159, 1971.

[100] W. Ma, C. Yang, X. Gong, K. Lee, and A. J. Heeger. Thermally stable, efficient polymer solar cells with nanoscale control of the interpenetrating network morphology. *Advanced Functional Materials*, 15(10):1617–1622, October 2005.

[101] O. Madelung, U. Rössler, and M. Schulz, editors. *Lead selenide (PbSe) energy gap and band structure*, volume 41C: Non-Tetrahedrally Bonded Elements and Binary Compounds. SpringerMaterials - The Landolt-Börnstein Database, 2010.

[102] O. Madelung, U. Rössler, and M. Schulz, editors. *Lead sulfide (PbS) energy gap*, volume 41C: Non-Tetrahedrally Bonded Elements and Binary Compounds. SpringerMaterials - The Landolt-Börnstein Database, 2010.

[103] O. Madelung, U. Rössler, and M. Schulz, editors. *Mercury telluride (HgTe) energy gap*, volume 41B: II-VI and I-VII Compounds; Semimagnetic Compounds. SpringerMaterials - The Landolt-Börnstein Database, 2010.

[104] M. P. Marder. *Condensed matter physics*. Wiley, Hoboken, NJ, 2. ed. edition, 2010.

[105] S. A. McDonald, G. Konstantatos, S. Zhang, P. W. Cyr, E. J. D. Klem, L. Levina, and E. H. Sargent. Solution-processed pbs quantum dot infrared photodetectors and photovoltaics. *Nat Mater*, 4(2):138–142, Feb. 2005.

[106] D. Mergel. Dünne ito-schichten als leitfhige, transparente elektroden. thin films of ito as conductive, transparent electrodes. *Vakuum in Forschung und Praxis*, 16(2):58–61, 2004.

[107] R. L. V. Metter, J. Beutel, and H. L. Kundel. *Handbook of Medical Imaging, Volume 1. Physics and Psychophysics*. SPIE Press, 2000.

[108] H. B. Michaelson. The work function of the elements and its periodicity. *Journal of Applied Physics*, 48(11):4729–4733, 1977.

[109] V. D. Mihailetchi, P. W. M. Blom, J. C. Hummelen, and M. T. Rispens. Cathode dependence of the open-circuit voltage of polymer:fullerene bulk heterojunction solar cells. *J. Appl. Phys.*, 94(10):6849–6854, 2003.

[110] V. D. Mihailetchi, J. K. J. van Duren, P. W. M. Blom, J. C. Hummelen, R. A. J. Janssen, J. M. Kroon, M. T. Rispens, W. J. H. Verhees, and M. M. Wienk. Electron transport in a methanofullerene. *Adv. Funct. Mater.*, 13(1):43–46, 2003.

[111] M. Morana, P. Koers, C. Waldauf, M. Koppe, D. Muehlbacher, P. Denk, M. Scharber, D. Waller, and C. Brabec. Organic field-effect devices as tool to characterize the bipolar transport in polymer-fullerene blends: The case of p3ht-pcbm. *Adv. Funct. Mater.*, 17:3274–3283, 2007.

[112] M. Morana, M. Wegscheider, A. Bonanni, N. Kopidakis, S. Shaheen, M. Scharber, Z. Zhu, D. Waller, R. Gaudiana, and C. Brabec. Bipolar charge transport in pcpdtbt-pcbm bulk-heterojunctions for photovoltaic applications. *Adv. Funct. Mater.*, 18:1757–1766, 2008.

[113] R. Morlotti, M. Nikl, M. Piazza, and C. Boragno. Intrinsic conversion efficiency of x-rays to light in gd2o2s : Tb3+ powder phosphors. *J. Lumin.*, 72-74:772 – 774, 1997. Luminescence and Optical Spectroscopy of Condensed Matter.

[114] D. Mühlbacher, M. Scharber, M. Morana, Z. Zhu, D. Waller, R. Gaudiana, and C. Brabec. High photovoltaic performance of a low-bandgap polymer. *Adv. Mater.*, 18:2884–2889, 2006.

[115] J. Nakamura, K. Murata, and K. Takahashi. Relation between carrier mobility and cell performance in bulk heterojunction solar cells consisting of soluble polythiophene and fullerene derivatives. *Applied Physics Letters*, 87(13):132105, 2005.

[116] K. K. Nanda, F. E. Kruis, H. Fissan, and S. N. Behera. Effective mass approximation for two extreme semiconductors: Band gap of pbs and cubr nanoparticles. *Journal of Applied Physics*, 95(9):5035–5041, 2004.

[117] D. Natali, M. Sampietro, M. Arca, and F. A. Denotti, C.and Devillanova. Wavelength-selective organic photodetectors for near-infrared applications based on novel neutral ditholenes. *Synth. Met.*, 137:1489–1490, 2003.

[118] G. Nootz, L. A. Padilha, L. Levina, V. Sukhovatkin, S. Webster, L. Brzozowski, E. H. Sargent, D. J. Hagan, and E. W. Van Stryland. Size dependence of carrier dynamics and carrier multiplication in pbs quantum dots. *Phys. Rev. B*, 83:155302, Apr 2011.

[119] D. C. Oertel, M. G. Bawendi, A. C. Arango, and V. Bulović. Photodetectors based on treated cdse quantum-dot films. *Applied Physics Letters*, 87(21):213505, 2005.

[120] R. Pacios. *Organic Photovoltaic Cells and Photodiodes based on Conjugated Polymers*. PhD thesis, Imperial College London, May 2003.

[121] F. Padinger, R. S. Rittberger, and N. S. Sariciftci. Effects of postproduction treatment on plastic solar cells. *Adv. Funct. Mater.*, 13:85–88, 2003.

[122] S. S. Pandey, W. Takashima, S. Nagamatsu, T. Endo, M. Rikukawa, and K. Kaneto. Regioregularity vs regiorandomness: Effect on photocarrier transport in poly(3-hexylthiophene). *Japanese Journal of Applied Physics*, 39(Part 2, No. 2A):L94–L97, 2000.

[123] R. Pätzold. *Organische Leuchtdioden auf flexiblen Substraten*. PhD thesis, Friedrich-Alexander-Universitt Erlangen-Nürnberg, 2005.

[124] R. Pätzold, A. Winnacker, D. Henseler, V. Cesari, and K. Heuser. Permeation rate measurements by electrical analysis of calcium corrosion. *Review of Scientific Instruments*, 74(12):5147–5150, 2003.

[125] J. Peet, M. L. Senatore, A. J. Heeger, and G. C. Bazan. The role of processing in the fabrication and optimization of plastic solar cells. *Adv. Mater.*, 21:1521–1527, 2009.

[126] E. Perzon, F. Zhang, M. Andersson, W. Mammo, O. Ingans, and M. R. Andersson. A conjugated polymer for near infrared optoelectronic applications. *Adv. Mater.*, 19:3308–3311, 2007.

[127] S. Pichler, T. Rauch, R. Seyrkammer, M. Böberl, S. F. Tedde, J. Fürst, M. V. Kovalenko, U. Lemmer, O. Hayden, and W. Heiss. Temperature dependent photoresponse from colloidal pbs quantum dot sensitized inorganic/organic hybrid photodiodes. *Applied Physics Letters*, 98(5):053304, 2011.

[128] J. J. H. Pijpers, R. Ulbricht, K. J. Tielrooij, A. Osherov, Y. Golan, C. Delerue, G. Allan, and M. Bonn. Assessment of carrier-multiplication efficiency in bulk pbse and pbs. *Nat. Phys.*, 5(11):811–814, Nov. 2009.

[129] M. Punke. *Organische Halbleiterbauelemente für mikrooptische Systeme*. PhD thesis, Universitt Karlsruhe, 2007.

[130] M. Punke, S. Valouch, S. W. Kettlitz, N. Christ, C. Gärtner, M. Gerken, and U. Lemmer. Dynamic characterization of organic bulk heterojunction photodetectors. *Applied Physics Letters*, 91(7):071118, 2007.

[131] T. Rauch, M. Böberl, M. Kovalenko, S. Tedde, U.Lemmer, J. Fürst, W. Heiss, and O. Hayden. Near infrared sensitivity of pbs quantum dot sensitized organic photodiodes. In *DPG-Spring Meeting, Berlin, Germany, February 25-29, 2008*, 2008.

[132] T. Rauch, M. Böberl, M. Kovalenko, S. Tedde, E. Zaus, U.Lemmer, J. Fürst, W.Heiss, and O. Hayden. Organic photodetectors sensitized with pbs nanocrystals for the near-infrared. In *SPIE Europe Photonics Europe, Strasbourg, France, April 7?10, 2008*, 2008.

[133] T. Rauch, M. Böberl, S. Tedde, J. Fürst, M. Kovalenko, G. Hesser, U.Lemmer, W.Heiss, and O. Hayden. Quantum dot sensitizers for near-infrared imaging with hybrid photodiodes. In *European Material Research Society Spring Meeting, Strasbourg, France, June 8-12, 2009*, 2009.

[134] T. Rauch, M. Boberl, S. F. Tedde, J. Furst, M. V. Kovalenko, G. Hesser, U. Lemmer, W. Heiss, and O. Hayden. Near-infrared imaging with quantum-dot-sensitized organic photodiodes. *Nat. Photon.*, 3(6):332–336, June 2009.

[135] T. Rauch, M. Boeberl, S. Tedde, J. Fuerst, M. Kovalenko, G. Hesser, U.Lemmer, W.Heiss, and O. Hayden. Near-infrared imaging with quantum dot sensitized organic photodiodes. In *2nd International Workshop on Semiconducting Nanoparticles, Duisburg, Germany, December 10-12, 2008*, 2008.

[136] T. Rauch, D. Henseler, P. Schilinsky, C. Waldauf, J.Hauch, and C. J. Brabec. Performance of organic photodetectors. In *4th IEEE Conference on Nanotechnology*, pages 632–634, 2004.

[137] T. Rauch, M. Sramek, S. F. Tedde, and O. Hayden. Infrarotdetektion mit organischen photodioden. In *Sensoren und Messsysteme*, page 166. Informationsgesellschaft im VDE (ITG) VDE/VDI-Gesellschaft Mess- und Automatisierungstechnik (GMA), VDE Verlage, 2010.

[138] T. Rauch, S. Tedde, and O. Hayden. Near-infrared imaging: hybrid photodiodes promise cost-effective short-wave ir imaging. *Laser Focus World*, pages 68–70, December 2009.

[139] P. Reiss, E. Couderc, J. De Girolamo, and A. Pron. Conjugated polymers/semiconductor nanocrystals hybrid materials-preparation, electrical transport properties and applications. *Nanoscale*, 3:446–489, 2011.

[140] A. Rogach, S. Kershaw, M. H. M. Burt, A. Kornowski, and H. W. A. Eychmüller. Colloidally prepared hgte nanocrystals with strong room temperature infra-red luminescence. *Adv. Mater.*, 11:552–55, 1999.

[141] A. L. Rogach, A. Eychmüller, S. G. Hickey, and S. V. Kershaw. Infrared-emitting colloidal nanocrystals: Synthesis, assembly, spectroscopy, and applications. *Small*, 3(4):536–557, 2007.

[142] E. H. Sargent. Infrared photovoltaics made by solution processing. *Nat Photon*, 3(6):325–331, June 2009.

[143] N. S. Sariciftci, L. Smilowitz, A. J. Heeger, and F. Wudl. Photoinduced electron transfer from a conducting polymer to buckminsterfullerene. *Science*, 258:1474–1476, 1992.

[144] R. D. Schaller, M. Sykora, Pietryga, J. M., and V. I. Klimov. Seven excitons at a cost of one: Redefining the limits for conversion efficiency of photons into charge carriers. *Nanoletters*, 6:424–429, 2006.

[145] M. C. Scharber, D. Mühlbacher, M. Koppe, P. Denk, C. Waldauf, A. Heeger, and C. Brabec. Design rules for donors in bulk-heterojunction solar cells - towards 10% energy-conversion efficiency. *Advanced Materials*, 18(6):789–794, 2006.

[146] P. Schilinsky, U. Asawapirom, U. Scherf, M. Biele, and C. J. Brabec. Influence of the molecular weight of poly(3-hexylthiophene) on the performance of bulk heterojunction solar cells. *Chemistry of Materials*, 17(8):2175–2180, 2005.

[147] P. Schilinsky, C. Waldauf, J. Hauch, and C. J. Brabec. Polymer photovoltaic detectors: progress and recent developments. *Thin Solid Films*, 451-452:105 – 108, 2004. Proceedings of Symposium D on Thin Film and Nano-Structured Materials for Photovoltaics, of the E-MRS 2003 Spring Conference.

[148] P. Schilinsky, C. Waldauf, J. Hauch, and C. J. Brabec. Simulation of light intensity dependent current characteristics of polymer solar cells. *Journal of Applied Physics*, 95(5):2816–2819, 2004.

[149] R. Schlaf, B. A. Parkinson, P. A. Lee, K. W. Nebesny, G. Jabbour, B. Kippelen, N. Peyghambarian, and N. R. Armstrong. Photoemission spectroscopy of lif coated al and pt electrodes. *Journal of Applied Physics*, 84(12):6729–6736, 1998.

[150] J. M. Schmitt, S. H. Xiang, and K. M. Yung. Differential absorption imaging with optical coherence tomography. *J. Opt. Soc. Am. A*, 15(9):2288–2296, 1998.

[151] B. Schrader and P. Rademacher. *Kurzes Lehrbuch der Organischen Chemie*, chapter 1. Grundlagen der Organischen Chemie, pages 1–66. DE GRUYTER, Germany, 2009.

[152] D. W. Schubert and T. Dunkel. Spin coating from a molecular point of view: its concentration regimes, influence of molar mass and distribution. *Mater. Res. Innovations*, 7:314–321, 2002.

[153] B. R. Sever, N. Kristianpoller, and F. C. Brown. F -center production in alkali halide crystals by monochromatic x-ray and ultraviolet radiation. *Phys. Rev. B*, 34(2):1257–1263, Jul 1986.

[154] E. V. Shevchenko, D. V. Talapin, H. Schnablegger, A. Kornowski, O. Festin, P. Svedlindh, M. Haase, and H. Weller. Study of nucleation and growth in the organometallic synthesis of magnetic alloy nanocrystals: The role of nucleation rate in size control of copt3 nanocrystals. *Journal of the American Chemical Society*, 125(30):9090–9101, 2003. PMID: 15369366.

[155] V. Shrotriya, G. Li, Y. Yao, C.-W. Chu, and Y. Yang. Transition metal oxides as the buffer layer for polymer photovoltaic cells. *Applied Physics Letters*, 88(7):073508, 2006.

[156] V. Shrotriya, G. Li, Y. Yao, C.-W. Chu, and Y. Yang. Transition metal oxides as the buffer layer for polymer photovoltaic cells. *Applied Physics Letters*, 88(7):073508–+, Feb. 2006.

[157] H. Sirringhaus, N. Tessler, and R. H. Friend. Integrated optoelectronic devices based on conjugated polymers. *Science*, 280(5370):1741–1744, 1998.

[158] H. Sirringhaus, N. Tessler, and R. H. Friend. Integrated optoelectronic devices based on conjugated polymers. *Science*, 280:1741–1744, 1998.

[159] D. N. Sitter, J. S. Goddard, and R. K. Ferrell. Method for the measurement of the modulation transfer function of sampled imaging systems from bar-target patterns. *Appl. Opt.*, 34(4):746–751, 1995.

[160] C. Soci, I.-W. Hwang, D. Moses, D. W. Z. Zhu, R. Gaudiana, C. Brabec, and A. Heeger. Photoconductivity of a low-bandgap conjugated polymer. *Adv. Funct. Mater.*, 17:632–636, 2007.

[161] M. Spahn. Flat detectors and their clinical applications. *European Radiology*, 15(9):1934–1947, Sept. 2005.

[162] W. P. Su, J. R. Schrieffer, and A. J. Heeger. Solitons in polyacetylene. *Phys. Rev. Lett.*, 42:1698–1701, Jun 1979.

[163] V. Sukhovatkin, S. Hinds, L. Brzozowski, and E. H. Sargent. Colloidal quantum-dot photodetectors exploiting multiexciton generation. *Science*, 324(5934):1542–1544, 2009.

[164] S. M. Sze and K. K. Ng. *Physics of Semiconductor Devices*, chapter Physics and Properties of Semiconductors - A Review, pages 5–75. John Wiley & Sons, Inc., 2006.

[165] S. M. Sze and K. K. Ng. *Physics of Semiconductor Devices*, chapter Photodetectors and Solar Cells, pages 663–742. John Wiley & Sons, Inc., 2006.

[166] S. M. Sze and K. K. Ng. *Physics of Semiconductor Devices*, chapter p-n Junctions, pages 77–133. John Wiley & Sons, Inc., 2006.

[167] S. M. Sze and K. K. Ng. *Physics of Semiconductor Devices*, chapter LEDs and Lasers, pages 599–662. John Wiley & Sons, Inc., 2006.

[168] S. M. Sze and K. K. Ng. *Physics of Semiconductor Devices*, chapter Metal-Insulator-Semiconductor Capacitors, pages 197–240. John Wiley & Sons, Inc., 2006.

[169] S. M. Sze and K. K. Ng. *Physics of Semiconductor Devices*, chapter Metal-Semiconductor Contacts, pages 134–196. John Wiley & Sons, Inc., 2006.

[170] S. M. Sze and K. K. Ng. *Physics of Semiconductor Devices*, chapter Sensors, pages 743–771. John Wiley & Sons, Inc., 2006.

[171] S. M. Sze and K. K. Ng, editors. *Physics of semiconductor devices*. Wiley-Interscience, Hoboken, N.J., 3. ed. edition, 2007. Description based on print version record. - Elektronische Ressource.

[172] K. Szendrei, F. Cordella, M. V. Kovalenko, M. Böberl, G. Hesser, M. Yarema, D. Jarzab, O. V. Mikhnenko, A. Gocalinska, M. Saba, F. Quochi, G. B. Andrea Mura, P. W. M. Blom, W. Heiss, and M. A. Loi. Solution-processable near-ir photodetectors based on electron transfer from pbs nanocrystals to fullerene derivatives. *Adv. Mater.*, 21:683–687, 2009.

[173] D. V. Talapin, J.-S. Lee, M. V. Kovalenko, and E. V. Shevchenko. Prospects of colloidal nanocrystals for electronic and optoelectronic applications. *Chem. Rev.*, 110(1):389–458, Jan. 2010.

[174] J. Tang and E. H. Sargent. Infrared colloidal quantum dots for photovoltaics: Fundamentals and recent progress. *Advanced Materials*, 23(1):12–29, 2011.

[175] J. Tbd. Reference. *Placeholder*, 2011.

[176] S. F. Tedde, J. Kern, T. Sterzl, J. Fürst, P. Lugli, and O. Hayden. Fully spray coated organic photodiodes. *Nanoletters*, 9:980–983, 2009.

[177] E. von Hauff, J. Parisi, and V. Dyakonov. Investigations of the effects of tempering and composition dependence on charge carrier field effect mobilities in polymer and fullerene films and blends. *Journal of Applied Physics*, 100(4):043702–+, Aug. 2006.

[178] C. Waldauf, M. Morana, P. Denk, P. Schilinsky, K. Coakley, S. A. Choulis, and C. J. Brabec. Highly efficient inverted organic photovoltaics using solution based titanium oxide as electron selective contact. *Applied Physics Letters*, 89(23):233517, 2006.

[179] C. Waldauf, M. C. Scharber, P. Schilinsky, J. A. Hauch, and C. J. Brabec. Physics of organic bulk heterojunction devices for photovoltaic applications. *Journal of Applied Physics*, 99(10):104503, 2006.

[180] C. Waldauf, P. Schilinsky, J. Hauch, and C. J. Brabec. Material and device concepts for organic photovoltaics: towards competitive efficiencies. *Thin Solid Films*, 451-452:503 – 507, 2004. Proceedings of Symposium D

on Thin Film and Nano-Structured Materials for Photovoltaics, of the E-MRS 2003 Spring Conference.

[181] Y. Wang, A. Suna, W. Mahler, and R. Kasowski. Pbs in polymers. from molecules to bulk solids. *The Journal of Chemical Physics*, 87(12):7315–7322, 1987.

[182] M. M. Wienk, M. G. R. Turbiez, M. P. Struijk, M. Fonrodona, and R. A. J. Janssen. Low-band gap poly(di-2-thienylthienopyrazine):fullerene solar cells. *Appl. Phys. Lett.*, 88:153511, 2006.

[183] F. W. Wise. Lead salt quantum dots:? the limit of strong quantum confinement. *Accounts of Chemical Research*, 33(11):773–780, 2000. PMID: 11087314.

[184] Y. Yao, Y. Liang, V. Shrotriya, S. Xiao, L. Yu, and Y. Yang. Plastic near-infrared photodetectors utilizing low band gap polymer. *Adv. Mater.*, 19:3979–3983, 2007.

[185] G. Yu, J. Gao, J. C. Hummelen, F. Wudl, and A. J. Heeger. Polymer photovoltaic cells: Enhanced efficiencies via anetwork of internal donor-acceptor heterojunctions. *Science*, 270:1789–1791, 1995.

[186] M. H. Zarghami, Y. Liu, M. Gibbs, E. Gebremichael, C. Webster, and M. Law. p-type pbse and pbs quantum dot solids prepared with short-chain acids and diacids. *ACS Nano*, 4(4):2475–2485, 2010. PMID: 20359235.

[187] E. S. Zaus. *Entwicklung und Charakterisierung von organischen Photodetektoren*. PhD thesis, Universitt Erlangen-Nürnberg, Universittsstrae. 4, 91054 Erlangen, 2007.

[188] A. Zen, J. Pflaum, S. Hirschmann, W. Zhuang, F. Jaiser, U. Asawapirom, J. P. Rabe, U. Scherf, and D. Neher. Effect of molecular weight and annealing of poly(3-hexylthiophene)s on the performance of organic field-effect transistors. *Adv. Funct. Mater.*, 14(8):757–764, 2004.

[189] S. Zhang, P. W. Cyr, S. A. McDonald, G. Konstantatos, and E. H. Sargent. Enhanced infrared photovoltaic efficiency in pbs nanocrystal/semiconducting polymer composites: 600-fold increase in maximum power output via control of the ligand barrier. *Applied Physics Letters*, 87(23):233101, 2005.

Acknowledgements

Let's come to those people, without whom this work would have never been possible. At the very beginning I would like to thank especially Dr. habil. Oliver Hayden as supervisor. Besides the knowledge and creativity, his ambition, curiosity and impatience kept working me hard on the results for the thesis and was pushing me to the limits. Many thanks to Dr. Jens Frst for giving me the opportunity to join such a nice team and the research field of organic optoelectronics.

Especially in the latest phase of this work, Professor Dr. Uli Lemmer was a great support and mentor to finalize the dissertation which took a while. Many thanks also to Sebastian Valouch and Siegfried Kettlitz at Professor Lemmer's institute who gave me the opportunity to perform fast transient optical measurements for my work.

I am very grateful to Professor Dr. Wolfgang Heiss from the University of Linz for providing excellent colloidal quantum dots in combination with fruitful analysis of obtained experimental data. Special thanks to Stefan Pichler for the photoluminescence measurements at low temperatures, Dr. Maxym Kovalenko for the outstanding synthesis of the quantum dots and Gunter Hesser for the high quality SEM and TEM characterization.

A very eminent support during my work was Dr. Michaela Bberl who was an essential help in the introduction to colloidal quantum dots and in the discussion for the design of the experiments and their results.

Dr. Sandro Tedde and Dr. Edgar Zaus made the years at Siemens Erlangen to an unforgettable experience because of the amicable and cooperative relationship on personal as well as on professional level.

A very special thank to Dr. Joachim Wecker for accepting me in his research group for organic electronics and for giving me this great chance.

Furthermore I want to thank all the other important supporters who were of great help for my dissertation: Johannes Kern, Jrgen Parzefall, Philip Nicklaus, Regina Pflaum, Eric Butkus, Tobias Sterzl, Line Domtchouang, Azrul Bin Sulamain, Samuele Lilliu, Daniele Parrotta, Antonio Castellano, Martina Dressler,

Jiaye Huang, Pierpaolo Lecis, Jan Matysik, Martin Miller, Mauro Morana, Josef Ott, Roberto Pacios, Nadege Pouatcha, Oliver Rudolph and Liu Yang.

At the very end I would like to express my gratitude to my parents Karl and Christine who already got nervous since their son took so long to get a reasonable job. Many thanks for the hosting and all the support during that long journey.

Finally, my wife Mona deserves the biggest part from the cake of gratitude. A lot of days she kept me away from everyday's duties to give me time and possibilities to work on my thesis. I am very grateful to haver her in my life.

To end, I want to apologize to my son Philipp and to my daughter Theresa, since they had to see their dad fewer times.